Louis Becke

Wild Life in Southern Seas

Louis Becke

Wild Life in Southern Seas

ISBN/EAN: 9783337003753

Printed in Europe, USA, Canada, Australia, Japan

Cover: Foto ©berggeist007 / pixelio.de

More available books at **www.hansebooks.com**

WILD LIFE IN SOUTHERN SEAS

Wild Life in Southern Seas

By Louis Becke

LONDON

T. FISHER UNWIN

1897

CONTENTS

vii

Orca Gladiator.

WE—a little girl of six, and myself—were seated upon a high, flat-topped, grassy headland of a lonely part of the northern coast of New South Wales, five miles from the old penal settlement of Port Macquarie. Three hundred feet below, the long Pacific rollers, unruffled by the faintest breath of air, swept in endless but surfless succession around a chain of black, isolated, and kelp-covered rocks that stood out from the shore at a distance of a cable-length or so. The tide was low, and some of the rocks raised their jagged, sun-dried summits perhaps six feet above the surface; others scarce a foot, so that each gentle swell as it came wavering shoreward poured over their faces in a creamy lather of foam; others again were fathoms below, and their thick garments of kelp and weed swayed to and fro unceasingly to the sweep of the ocean roll above

them. And in and about the rocks, and hovering over the white gleam of sandy bottom that,
like a great table of ivory, lay between them
and the cliff-bound shore, swam droves of
bright, pink-coloured schnapper, and great,
lazily moving blue-fish. Half a mile away a
swarm of white gulls floated motionless upon
the blue expanse ; upon the time-worn fore-
shore boulders beneath us stood lines and groups
of black divers, with wings outspread in solemn
silence, gazing seaward.

We had climbed the headland to look for
whales ; for it was the month of October, when
the great schools of humpbacks and finbacks
were travelling southward to colder seas from
their breeding grounds among the Bampton
Shoals, nine hundred miles away, north-east.
For three weeks they had been passing south,
sometimes far out from the land, sometimes
within a mile of the shore—hundreds of
thousands of pounds' worth of rich blubber,
with never a whaleship nor whaleboat's crew
within two thousand miles ; for the brave old
days of Australian whaling enterprise died full
thirty years ago.

At last, a mile or so away, a jet of smoky

spray, and then another and another! Five humpbacks—two cows, two calves, and a bull —only a small " pod "—that is, a school. Nearer and nearer they came, their huge, black humps gleaming brightly in the dazzling sunlight as they rose to spout. A hundred yards in front, the old bull rolls lazily along, "sounding " but rarely, for the sea is full of squid, and he and his convoy, with drooping lower jaws, suck in the lovely morsels in countless swarms.

Six weeks before, as they had rolled and spouted northwards to the great lagoons of the Bampton and Bellona Reefs, they had passed within, perhaps, a hundred yards of the headland upon which we sat. Perhaps, too, a fierce " north-easter " blew, and the chain of rocks that was now so gently laved by the murmuring waves was smothered in the wild turmoil of a roaring surf, and the great bull, although his huge, corrugated belly itched sorely from the thick growth of inch-long barnacles that had so tormented him of late, spouted regretfully and headed seaward again—even he could not scratch his giant frame in such a surf as that. But to-day it was different ; and now he could

enjoy that long-delayed pleasure of dragging
his great body over the rough surfaces of the
submerged rocks, and tearing those dreadfully
irritating barnacles off his twenty-five feet of
grey-white ridgy stomach. For, suddenly, he
raised his vast head, and then " sounded,"
straight on end, and the child by my side
gave a gasp of wondering terror as she saw
his mighty tail rise a good ten feet in air
and then slowly vanish beneath the sea.

On went the cows and calves, apparently
taking no heed of father's sudden dive shoreward. He would soon be back, they knew, as
soon as the poor fellow had rid himself of those
tormenting barnacles ; and so with diminished
speed they kept in southwards towards Camden
Haven. But just as the great bull came bursting through the blue depths into the greeny
hue of six fathoms of water, we saw between
him and the " pod " two small jets, like spurts
of steam, shoot up from the water between him
and his convoy ; and in another second the
cows and calves had sounded in deadliest terror,
and were rushing seaward, two thousand feet
below. For they knew that out there in the
depths lay their only hope of safety from their

,

dreaded and invincible enemies, the "killers" and "threshers" of the South Pacific—the murderous, savage cetacean pirates that lie in wait for the returning "pods" as they travel southwards to the colder seas of Tasmania. As the great humpback reached the chain of rocks, and had begun to scratch, his foes had advanced silently but swiftly towards him. Before them swam their equally fierce and dreaded ally, *Alopias Vulpes*, the "thresher," or fox-shark. But, before I tell of that noble fight of giants, which for nearly two hours we gazed at on that October morning from the lonely headland, let me say something about *Alopias Vulpes* and his fellow pirate, *Orca Gladiator*, the "killer."

First of all, then, as to the "thresher." He is a shark, pure and simple, and takes his name from his enormous, scythe-like, bony tail, which forms two-thirds of his length. His mouth is but small, and whales have little to fear from that, but dread the terrible knife-like sweep and downward slash of his tail ; for each stroke cuts through the tough skin and sinks deep into the blubber. Such is the "thresher," and in every drove of "killers" there is always one thresher, sometimes two.

The "killer" is actually a whale, for he is warm-blooded and rises to the surface to spout, which he does in a manner that has often led to his being mistaken for a humpback, or finback whale. He is distinguishable only from the grampus by his mouth, which has teeth—and terrible teeth—in both jaws : the grampus has teeth in his lower jaw only. When he (the grampus) is a baby he has teeth in both jaws, but those of the upper jaw are shed and fall out when he is about half grown. The killer has teeth in both jaws, as many a poor humpback and finback has found out to his cost, for the fierce creature does justice to his name—*Orca Gladiator.*

The killers have a business, and they never neglect it. It is the business of whale catching and killing. They are the bull-dog pirates of the deep sea, and on the coast of Australia their headquarters are at Twofold Bay. Sometimes, but not often, they have been known to attack the monarch of the ocean, the sperm whale. But they generally leave him alone. He is too big, too powerful, and his great eight-inch teeth and fierce spirit render him a dangerous customer to tackle. But with the right whale, the humpback, and the seventy-foot flying finback, the killers can work their cruel will.

And now to the fight we saw.

.   .   .   .   .

For about ten minutes or so the great humpback dragged his monstrous fifty feet of flesh and blubber across the tops of the submerged rocks, raising sometimes his vast head and sometimes his mighty flukes out of the water, as with all the weight of his giant body he rubbed, and scraped, and scratched his itching belly against the surface of the rocks. Suddenly, a long, slender, greyish object swept like lightning upon him, and the thresher buried his teeth in the loose skin of his " small "—that is, about fifteen or sixteen feet from his tail. And at the same moment, with savage puffs of spray shooting high from their blow-holes, the two killers darted at his head and seized him by the jaws. In ten seconds there was nought to be seen but a maddened whirl and seeth of foam, as the unfortunate victim sought to escape seaward. Well did he know that in such shallow water—there was but five or six fathoms—he could not sound far below into ocean's depths, and, carrying his foes with him, compel them to rise for air. Fifteen, perhaps twenty minutes exhausts the air supply of a killer ; a whale can remain below the surface for sixty. But he made a bold attempt.

Raising his enormous head high in air, and giving it a mighty shake, he freed himself from one of the killers, whose body, twenty feet in length, he hurled from him as if it were a minnow ; but the other, with his cruel teeth buried bull-dog fashion in his thick lips, hung on with savage tenacity. And down upon his " small " the thresher, with his teeth gripping the loose, tough, and wrinkled skin, upreared his lengthy form, and brought his awful scythe-like tail down upon the victim's back, with a smack that could be heard half a mile away. It cut, and then, as the whale rolled in his agony from the blow, a broad, white streak of blubber oozed through the severed skin. Before he could gather his strength for that seaward rush, which meant life, the thrown-off killer was back, and had seized him again by his starboard lip. Too late ! he could not sound and could not flee, and the poor, worried animal seemed to know it, for suddenly he lay quiet, while the bulldogs shook him and the thresher dealt him steady but fearful blows upon his broad expanse of shining back.

" Oh, the poor whale ! " said my little companion, as she shudderingly clutched my arm. " Look at that ! "

The killer fastened to the left jaw of the helpless, floating monster, raising his square white and black head about a foot or two out of the water, gave it that quick jerk one sees a fox-terrier give to a rat, and brought away in his jaws a piece of lip about a yard long—a thick strip of bloody white and red. And, as a terrier throws a rat backward and upward, so did the killer throw away the gory mass ; it fell with a heavy splash upon the water some fathoms away. Then with a mighty leap the wretched whale sprang clear out of the water, standing for a moment or two straight up and down, and as he swung his body round in falling, we saw the blood pouring from his jaws in a stream. He fell upon his back with a terrific splash of foam, and for a few seconds was out of sight ; again he raised his head—the killers were both fastened to his lips again, tearing off the blubbery flesh in monstrous strips.   Once, as he wallowed in his agony, he opened his vasty jaw, and ere he could close his mouth one of his foes thrust in his bull-dog head and sought to tear away a piece of his great tongue.   And then came such a crashing and splashing and bewildering leaping of foam, and his tail upreared itself and swept round and round in all directions, and then struck the water a blow that sounded like a thunderclap.

"Look," said the child again, "there are more of those cruel killers coming ; see, there they are, just below us !   Oh ! how I hate them !"

Fifty feet away from the persecuted hump-back, and sailing round and round in the green water beyond the rocks, were five sharks.   They had smelt the blood of the battle, and were waiting till they could join in, and, while the killers forced their heads into the humpback's mouth and tore out his toothsome tongue, feed upon the quivering mass of blubber and rend him in pieces from his head down to his "small."

The unfortunate animal was now becoming

rapidly exhausted, and although he still struck the water resounding blows with his tail, he was convulsed with pain and terror, and swam slowly round and round in a circle, spouting feebly, and rolling from side to side in a vain effort to shake off the killers, and find his way to the open sea. Then, as if wearied with their attempts to get at his tongue, the two destroyers suddenly let go their hold and swam away some twenty yards or so ; and the thresher, too, although he still lay alongside, ceased his fearful blows and let his long, narrow, and tapering body lie motionless upon the water, and the five grey sharks drew nearer and nearer. But the killers had not left him, for after spouting once or twice, they slewed round and came at the prey with a savage rush, and, leaping bodily out of the water, flung themselves upon his back time and time again with the most cruel and extraordinary pertinacity. And so, at last, there he lay, his monstrous head and thirty feet of his back raised high out of the water, and the white seethe of foam in which his colossal frame writhed and shuddered in deadly torment was tinged deeply with a bloodied red. Better far would have been for him the swift, death-dealing stroke of the whaler's lance, or the dreadful "squish" of the bursting bomb as it entered his vitals, and put an end to him at once, than endure such tortures as now were his. But, presently, gathering his strength for one final effort, one last spout slowly curled out, he lowered his head, raised his tail, and dashed headlong seaward. And like demons from the pit the two killers followed him down. They knew that for a mile out the water was too shallow for him to get away from them. Behind, the five sharks swept in swift pursuit ; ahead of all *Alopias Vulpes* cleft the water with sharp vicious "tweeps" of his long tail.

Five, perhaps six, minutes passed, and then, with a roaring burst of foam, and spouting quickly, he raised his immense form half out of the water and, supporting himself upon his tail, spun round and round. Twice his cave-like mouth opened and shut, and as he beat the sea into froth and spume around, a strange, awe-inspiring sound accompanied his last spout ; for the sharks were at him below, tearing and riving out mouthfuls of blubber, and the killers

had dragged out his tongue. One last shudder-
ing gasp, and the now unconscious creature
sank backward, and describing a circle in his
final " flurry," rolled over, " fin up," and gave
up his greasy ghost.

Green Dots of the Empire: The Ellice Group.

DOTS only. And if the ship that carries you is running past them in the night, with the steady force of the south-east trades filling her canvas, you would never know that land lay within a few miles, save for the flashing of lights along the low sandy beaches or, mayhap, the dulled roar of the beating surf thrashing the reef on the windward side of the island. This, of course, implies that when ships pass in the night they do so on the lee-side. It is not a safe thing for even a daring trading schooner to have a long, long stretch of low-lying reef-encircled islets for a lee ; for sometimes *Matagi toe lau* (as the brown-skinned people call the trade wind) is apt, a few hours before dawn, to lull itself to slumber for a space, till the sun, bursting from the ocean, wakes it to life again.

14

And should the schooner have drifted down upon the land with the stealthy westerly current there is no such thing as trusting even to good ground tackle on the weather side of an Ellice Group atoll. Did the ocean slumber too, and the black ledges of the windward reef were laved but by the gentlest movement of the water, there would be no anchorage, unless the ship were loaded with a cable long enough ; and ere the sun has dried the dews of the night on the coconuts the merry trade wind pipes up again, the smooth surface of the ocean swells and undulates, the rollers sweep in from the eastward and charge wildly against the black wall of coral rock, smothering it in a maddened tumble of froth and foam, the while the smoky sea-spume is carried on high to fall in drenching showers upon the first line of coco-palms and *puka* scrub growing down close to the iron-bound shore. And of the eight islands of the Ellice Group all are alike in this respect—a wild tumultuous surf for ever beats upon the weather shore even under the influence of the ordinary trade wind ; and, on the lee, there lies a sea as placid and motionless as a mountain lake.

.　　.　　.　　.　　.

Four years ago the Gilbert and Kingsmill Groups (known collectively as the Line Islands) and the Ellice Group were annexed by Great Britain ; and although people in Australia hear and read a good deal about the Gilberts and Kingsmills by reason of their being the location of the newly-appointed British Resident and Deputy-Commissioner for the Western Pacific, seldom is anything heard about or told of the almost equally important Ellice Group. The reason for this is not far to seek. The Line Islanders — fierce, turbulent, and war-loving people, island hating island with the same savage animosity that characterised the Highland clans of the thirteenth century—are a difficult race to govern, and although the London Missionary Society has done much good, the Resident has his work cut out to prevent the people of his sixteen islands shooting and cutting each other's throats as they did in the good old days. For when Captain Davis, of Her Majesty's ship *Royalist*, hoisted the English flag, he sternly intimated that there was to be no more fighting, and later on the High Commissioner, Sir John Thurston, in the *Rapid*, made them disarm ;

but scarce had the smoke from the steamer's funnel vanished from the horizon than the old leaven worked, and rifles, carefully hidden away from the naval men, were brought forth from their concealment and put to use. And so every few months or so the Australian newspapers notify that " there has been fresh trouble in the Gilbert Group." However, all this will be a thing of the past in another year or two, although it is safe to predict that it will be long ere the Gilbert Islander—man or woman —gives up the manufacture and use of sharks' teeth swords and daggers. And as these weapons are not necessarily fatal, and are time-honoured arguments for settling public and family differences, perhaps it will be as well for the High Commissioner to let them possess the means of letting out in a moderate degree some of their quick, hot blood.

But the people of the Ellice Group show the other side of the picture, and their calm, placid existence, undisturbed except by a family quarrel, explains why—saving the visit of a surveying ship—no men-of-war steam up to the anchorages outside the reefs, or into the lagoons, and hold courts of inquiry into native outbreaks or

3

private shootings.   The Ellice Islanders never
fight, for they have a horror of bloodshed, and
except for a few fowling-pieces used for shooting
pigeons, there are no firearms in the group—
save those in the possession of the white
traders.

.   .   .   .   .

Six hundred miles from Samoa, sailing north-
westerly, the first of the group, Sophia Island, is
sighted.   It is the south-easterly outlier of the
Ellices, and is the only one of sufficient height
to be seen from the vessel's deck at a distance of
twenty miles.   Until a few years ago it was uninhabited, although the people of the next island,
Nukulaelae, say that "in the old, old time many
people lived there."   It is about three and a half
miles in circumference, has but few coconuts
growing upon it, and would have remained
untenanted in its loneliness to this day but for
the discovery of a fairly valuable deposit of
guano.   Then it was taken possession of by
an enterprising American store-keeper in Samoa
named Moors, who landed native labourers and
worked, and is still working, the deposit.   The
old native name of this spot is Ulakita—a name,
by the way, that is almost unknown even to the

local traders in the Ellice Group, and the present generation of natives.

Eighty or ninety miles away is Nukulaelae, a cluster of thirteen low-lying islets, forming a perfect atoll, and enclosing with a passageless and continuous reef a lagoon five miles in length by three in width. This narrow belt of land— in no case is any one of the islets over a mile in width—is densely covered with coconuts, and, seen from the ship, presents an enchanting appearance of the brightest green, accentuated on the westerly or lee shore by beaches of the most dazzling white. Thirty years ago Nukulaelae had a population of four hundred natives.

Then one day, in 1866, there came along two strange vessels, a barque and a brig, and hove-to close to the reef, and in a few hours nearly two hundred of the unfortunate, unsuspecting, and amiable natives were seized and taken on board by the Peruvian cut-throats and kidnappers that had swept down upon them, and, with other companions in misery, torn from their island homes, taken away to slavery in the guano pits of the Chincha Islands, on the coast of South America. Of the Nukulaelae people none but two ever returned—they all perished miserably

under their cruel taskmasters on the gloomy
Chinchas. In 1873 it was the writer's lot to
meet, in the Caroline Islands, with one of the
two survivors of this dreadful outrage. By
some means he had escaped in an English guano
ship to Liverpool, and then, after years of
wandering in American whalers among the
islands of the Pacific, he settled down among
the natives of Las Matelotas, in the Carolines,
thousands of miles away from his birthplace ;
and although sorely tempted to accept the offer
made to him by our captain of a passage to
Nukulaelae, the Matelotas people refused to let
him go, as he had married a girl of the island
and had a family. (*Apropos* of these Peruvian
slavers, it may be mentioned that a few months
after their visit to Nukulaelae, joined by another
barque, they made a similar descent upon the
people of Rapa-nui—the mysterious Easter
Island—and secured three hundred and ten
victims.) At present the population of Nukulaelae is about one hundred and fifty, all of
whom are Christians. Like all the other islands
of *this* group, the population is showing a slow
but certain increase.

Within a few hours' sail lies Funafuti, an

extensive chain of some thirty-four or thirty-five islands similar in appearance to the islets of Nukulaelae, but enclosing a noble lagoon, entrance to which is given by good passages both on the south-west and north-west sides. The Russian navigator Kotzebue sailed his frigate through Funafuti Lagoon from one end to the other with a strong breeze blowing, and found, what trading vessels to-day know well, that unless a vessel is making something like eight knots it is almost impossible to stem the fierce current that sweeps through the passages at half-tide. But once well within the lagoon, and away from the trend of the passage current, there is room for half a dozen or more battleships in which to manœuvre. About six miles from the south-west entrance the ship may drop anchor off the main island of the chain; and here the native settlement is situated. Fifty years ago nearly every island in the lagoon supported a population ; to-day there are but four or five hundred natives all told, all of whom live on the island from which the whole group takes its general name, Funafuti.

The natives are a hospitable, good-tempered, and intelligent lot, and express themselves as

being delighted to be included as British
subjects. And there can be but little doubt
that in a few years, once assured of the good
intentions of the English authorities to them,
they will agree to lease out to traders and copra-
buyers the long stretch of dense but narrow sea-
girt coconut forests that form the southern
boundary of the lagoon. At present, and, indeed, for the past forty years, some millions of
coconut palms are there allowed to fruiten and
literally cover the ground with coconuts from
year to year without the natives gathering more
than will provide them with their few wants in
the way of clothing, tobacco, etc., which they
purchase from the one or two resident traders.
Time after time have the people been approached
by white agents of trading firms—notably in
years past by Godeffroy's of Hamburg—on the
subject of leasing one particularly noble island,
named Funafala, for the purpose of making the
coconuts into copra. Liberal terms—and for a
South Sea trading firm to offer liberal terms to
natives shows the value of the concessions
sought—were offered, but the Funafutans would
have none of the white men on Funafala. A
solitary trader or so they would tolerate in the

only village, but no body of strange, dissolute foreigners would they have to live among them, accompanied by wild people from the Gilbert Islands, who fought with sharks'-teeth swords among themselves, and got madly drunk on toddy every few days. And so the trading firms retired discomfited, and the coconuts rotted away quietly in millions, and the rotting thereof troubled the careless owners not a whit. Time was when there were three thousand people to eat them, and, save for a cask of coconut oil sold now and then to some whaleship, white men visited them but at long intervals. But things are different now, and even these tiny spots that dot the broad bosom of the blue Pacific are sought out to appease the earth-hunger of the men of the civilised world. Yet not, be it said, altogether for their coconuts' money value, but because of the new Pacific cable that is soon to be ; for among these equatorial isles it is to be laid, thousands of fathoms deep, and no Power but England must possess a foot of soil in the mid-Pacific that would serve an enemy as a lair whence to issue and seize upon any of the islands that break the cable's length

Take Funafuti and its people as a fair type of the other islands of the group, save Nui—of which more anon.　Sixty or seventy years ago, so the American whaleship captains of those days said, there were three thousand people in the thirty and odd islets.　Then, for the next thirty years, unknown and terrible diseases, introduced by the white men, ravaged not Funafuti alone, but the whole group, and where there were once thousands, only hundreds could be counted ; and until about 1860 it looked as if the total extinction of the whole race was but a matter of another decade.　But, fortunately, such was not the case.　In 1870 the writer counted 160 people ; in 1882 they had increased to nearly 200 ; and now, through better means of intercourse with the people of the other islands of the group, which has brought about a consequent and rapid inter-marriage, the people of Funafuti number over 500, and show a gradual but steady increase.

Oaitupu (literally " the fountain of water ") is, although nearly the smallest, the most thickly populated of all the Ellices.　It has no lagoon accessible from the sea, and even landing is not always easy.　Here, although the soil is better than that of the other islands, and the natives have taro, bananas, and pumpkins to vary the monotonous diet of coconut and fish obtaining elsewhere in the Ellices, they are very subject to that species of eczema known as *tinea desquamans* (locally it is called " lafa "). While not incapacitating them from labour, or affecting their stamina or physique, it gives the subject a most unpleasant and disgusting appearance. It is, however, often curable by a residence in a colder climate, such as New Zealand.

Nui, the island alluded to as possessing distinct and peculiar racial characteristics from the others, has a population of about six hundred. Unlike their neighbours, both to the north and south, whose language, customs, and traditions have a purely Samoan basis, the people of Nui are plainly the descendants of some wandering or drifted voyagers from the Gilbert Group, the inhabitants of which they resemble in language, customs, appearance, and demeanour. From what particular island the original people of Nui came is a mystery. There are no really reliable traditions of the present race that can throw any light on the matter. So far as they know they were always " Tafitos "—namely, people from the Gilberts ; but how they came to be on Nui they cannot tell. To show the sharp line of racial distinction between the natives of Nui and those of the surrounding islands, it may be mentioned that the translations of the Old and New Testaments, published by the Boston Board of Missions for the use of the people of the Gilbert Islands, are used by the natives of Nui, while in every other island of the Ellice Group the Samoan version is alone understood and read. And although they can communicate with the inhabitants of the rest of the group in a peculiar, bastard patois, and of late years have intermarried with them, they always will be Gilbert Islanders, and preserve their vernacular and other racial characteristics.

Nanomaga, the Hudson Island of Commodore Wilkes, is the smallest of the group. It is barely a mile and a half long, and not one in width, yet supports a population of six hundred people. The writer, who in 1870 spent a year on the island, can bear testimony to the kindly nature and honesty of its people. During all the time he lived there as agent for Messrs. Tom De Wolf and Co., of Liverpool, he never had as much as a scrap of tobacco stolen from him, although his trade goods were piled up indiscriminately on the floor of his house, which had neither doors, locks, nor a bolt of any kind. In this, however, the Nanomagans are peculiar — the other islanders are not so particular.

The last of the group is Nanomea, a fine island, or rather two islands connected by a reef dry at very low tides. The people of Nanomea have long been known in the Pacific for their great size and muscular development. Indeed, the Rev. J. S. Whitmee, of the L.M.S., considers them a race of giants, and believes " that nine out of ten would measure six feet or more high, and their breadth is proportionate to their height." This, however, since their adoption of clothing is not so noticeable. However, they certainly are a fine race, and almost free from *tinea desquamans*. There were, last year, 830 people on the two islands, Nanomea and Lakena.

The group suffers but seldom from droughts or hurricanes, although the terrible drought experienced in the near-to Gilbert Group in 1892 also affected the Ellices, and during

1893–4 Nanomea and Nanomaga presented a parched-up appearance. A heavy blow in 1890 also did terrible havoc among the coconuts, which thus had not the strength to bear up against the drought.

The whole group of the nine islands or sub-groups lies between lat. 5.35 deg. and 11.20 deg. S., and between 171 and 176 deg. W. longitude.

The Tia Kau.

FOUR miles north-west from Nanomaga, a tiny isle of the lately annexed Ellice Group in the South Pacific, lies a great " patch " of submerged coral, called Tia Kau—the best fishing ground in all the wide South Sea, except, perhaps, the atolls of Arrecifos and Christmas Island, in the North Pacific. Thirty years ago, when the smoke and glare from many a whaler's try-pots lit up the darkness of the ocean night from the Kermadecs to the far Pelews, the Tia Kau was known to many a sailor and wandering trader. But now, since the whaling industry died, and the trading vessels are few and far between, the place is scarcely even known by name.

.   .   .   .   .

A hot, steamy mist lies low upon the glassy surface of the sleeping sea encompassing Nano-

maga, and the lazily swelling rollers as they
rise to the lip of the reef have scarce strength
enough to wash over its flat, weedy ledges
into the lagoon beyond.　For since early morn
the wind had died away ; and the brown-
skinned people of the little reef-girt island,
when they rose from their slumbers and looked
out upon the dew-soaked trees, and heard the
moan of the distant breakers away on Tia Kau,
said to one another that the day would be calm
and hot till the sun was high and the wind came.
And, as your true South Sea Islander dreads the
blistering rays of the torrid sun as much as he
does the stinging cold, each man lay down again
upon his mat and smoked his pipe or cigarette,
and waited for the wind to come.

　　Along the silent and deserted beach long lines
of coco-palms, which slope seaward to the trades,
hang their drooping, languid plumes high above
the shallow margin of the lagoon, which swishes
and laps in gentle wavelets along the yellow
sand.　A shoal of pale grey mullet swim close
inshore, for out beyond in the deepening green
flit the quick shadows of the ever-preying frigate
birds that watch the waters from above.

　　'Tis roasting hot indeed.　As the mist begins

to lift, the steely ocean gleam pains the eye like a vast sheet of molten lead, and the white stretch of sand above high-water mark in front of the native village seems to throb and quiver and waver to and fro ; the mat coverings of the long row of slender canoes further down crackle and warp and swell upward.

Presently the one white trader on the little island comes to the doorway of his house and looks out. Not a living thing to be seen, except, far out beyond the reef, where the huge bodies of two blackfish lie motionless upon the water, sunning themselves ; and just above his head, and sitting on its perch, a tame frigate-bird, whose fierce eye looks upward and outward at the blazing sun.

" What a terror of a day ! " mutters the trader to himself, as he drinks his morning coffee, and then lazily sinks into a cane lounge on his verandah. He, too, will go to sleep until the breeze springs up, or some inconsiderate customer comes to buy tobacco, or tell him the local gossip.

In and about the village—which is a little further back from the trader's house — the silence of the morning heat reigns supreme.

The early meal of fish and taro has been eaten, and every one is lying down, for the smooth white pebbles of sea-worn coral that cover the ground around the high-roofed houses of pandanus thatch are hot even to the native foot, though here and there may be a cool strip of darkened shade from the overhanging branch of palm or breadfruit tree. Look through the open doorway of a house. There they lie, the brown-skinned lazy people, upon the cool matted floor, each one with a wooden *aluga*, or bamboo pillow, under his or her head, with their long black tresses of hair lying loosely uncoiled about the shoulders. Only three people are in this house, a big reddish-brown skinned man, a middle-aged woman, and a young girl. The man's and woman's heads are on the one pillow; between them lies the mutual pipe smoked out in connubial amity; the girl lies over in the corner beside a heap of young drinking coconuts and a basket of taro and fish, her slender figure clothed in nought but a thick girdle of fine pandanus leaf. She, too, has been smoking, for in her little hand is the half of a cigarette.

A wandering pig, attracted by the smell of

food, trots slowly to the door, and stands
eyeing the basket. His sleepy grunt betrays
him, and awakens the girl, who flings her
bamboo pillow at his head with a muttered
curse ; and, crawling over to where her sleeping
parents lie, she pillows her head upon her
mother's naked thigh, and falls asleep again.

Another hour passes, and then a faint breath
moves and sways and rustles the drooping
palms around the village, and the girl awakes.
Had she been dreaming, or did she hear a far-
away curious sound—a mingling of sharp,
whistling notes and hoarse, deep gutturals, such
as one may hear when a flock of terns and
boobies are darting down upon their prey ?
Tossing back her black mane of hair, she
bends her head seaward and listens intently,
and then, rising, goes to the open door, and
looks out upon the shimmering blue. The
white man, too, has heard, and she sees him
running to the village. The dulled, sleepy look
in her big eyes vanishes, and darting over to
her slumbering father, she slaps his brawny arm.

"*Ala ! Ala !* awake, my father. There be a
flock of *gogo* crying loudly, and the white man
is running hither."

4

The big man springs to his feet, followed by his wife, and in a moment the whole village is awake, and the men run beachward to their canoes ; for the flock of *gogo* means that a shoal of bonito, perhaps twenty thousand or more, are passing the island on their way to Tia Kau.

Before the men, laden with their fishing tackle, have reached the canoes, the village children are there, throwing off the coverings of mats in readiness for launching, and then, with a merry clamour of voices, the slender craft are lifted up and carried down to the water's edge. The white man, too, goes with them in one Muliao's canoe, and the women laugh and wish him luck as they see him strip to the waist like one of their own people, and show a skin almost as brown.

Over the reef they go, thirty or more canoes, paddling to the west.   There, a mile beyond, is a vast flock of *gogo*—a small, sooty tern—the density of whose swaying cloud is mingled with the snowy white of gulls.   How they flutter, and turn, and dive, and soar aloft to dive again, feasting upon the shining baby *kanae*, or mullet, that seek to escape from the ravenous jaws of the bonito, whose way across the sea is marked by a wide streak of bubbling, hissing foam !

Meanwhile, as the canoes fly in pursuit, one man in each busies himself by hurriedly preparing his fellows' tackle, which is both for rod and deep-sea fishing. Lying side by side upon the *ama*, or outrigger grating, are four rods. And such rods ! twelve to fourteen feet in one piece, eight inches in circumference at the base, and tapering to an inch at the point. But big and clumsy as they look, they are light, tough, and springy. The line is of two-stranded *fau* (hibiscus bark), and is not quite as long as the rod itself; the shank of the hook is of pearl-shell, gleaming and iridescent as polished opal, and the upward curving piece that forms the barbless point is cunningly lashed to the heel of the shank with fine banana fibre. In length these hooks range from one to three inches, and at the lashing of the point and shank are two tiny scarlet feathers of the parrokeet. Lying beside the rods are the thick, neatly curled lines for deep-sea work. But just now these are not wanted.

And as the canoes draw near the whirling,

shrilly-crying birds, the water becomes a wild,
seething swirl of froth and foam, for the bonito
are travelling swiftly onward, snapping and
leaping at the persecuted *kanae*, and their tens
of thousands of bodies of shining blue and silver
sparkle brightly in the sun.   And then with a
wild shout of glee the leading canoes shoot into
the fray, quickly followed by the others.

" *Tu ! Tu !* " (" Stand up, stand ! ") cry the
paddlers amidships, and in an instant the men
seated for'ard and aft drop their paddles, seize
their rods, and each man bracing his right leg
against the rounded thwart on which he has
been sitting, swings his bright, baitless hook
into the whirl below. Almost ere it touches
the water a fish leaps to it, the tough rod of
pua quivers and trembles, the fisher grunts, and
then with a strong, swift, and steady sweep of
his naked arms, and a triumphant cry of
" *Maté !* " (" Struck ") the first *atu*, ten pounds
of sheeny blue and polished silver, is swung
clear of the water and dropped into the canoe,
where he kicks and struggles among the
paddlers' feet. In another minute every other
canoe is hard at work, and the loud shouts and
cries of the excited natives add to the din of

the wheeling birds and the splashing of the water and the furious kicking and thumping against the frail, resonant sides of the canoes, as fish after fish is swept upward and outward, and dropped struggling into the bottom, among its bleeding and quivering fellows.

Around the largest canoe, from which six natives fish, is the wildest boil and bubble of all, for the cunning crew have hung from a bended stick over the side a bright piece of mother-of-pearl, and at this the hungry fish leap fiercely. How they swarm and "ring" round the canoe like a mob of frightened cattle upon some wide Australian plain, who smell their deadly enemy—a wild black! Not that the bonito are frightened; they are simply mad for the shining hooks, which look so like young and tender half-grown flying-fish.

But still on and on the main body go, and the canoes go with them, steadily on to the Tia Kau, although now each man has taken perhaps twenty or thirty fish from eight to ten pounds in weight; and the paddlers' arms are growing weary. Already the white man is tired, and is sitting down, smoking his pipe, and watching the moving cloud of birds above. And yet his comrades swing their rods, and add fish after fish to the quivering heap below. Time enough for them to smoke, they think, when the fish are gone—and then, suddenly, with an almost noiseless " flur-r-r ! " they *are* gone, and the white man laughs ; he knows that there will be no more *atu* to-day. For there, swimming swiftly to and fro upon the now quiet surface are half a dozen *pala*, the dreaded foe of the bonito for all time.

The canoes come to a dead stop ; the shoal of *atu* have dived perhaps a hundred fathoms deep, and will be seen no more for many an hour. And so the natives sit down and smoke their pipes, and hurl reproaches and curses at the *pala* for spoiling sport.

" Why grumble, Muliao ? " asks the white man of his friend. " See, already the canoes are weighted down with fish. But yet let us catch one of these devils before we return to the shore."

" *Meitake !* Aye, that shall we, though who careth to eat of *pala* when bonito is to his hand ? But yet to punish these greedy devils for coming here——" and Muliao takes from the outrigger a coil of stout three-stranded line,

which he makes into a running bowline and
hangs over the side of the canoe from the end
of his rod, while another man picks up a small
bonito, passes a line through its gills, and then
throws it far out upon the water only to draw
it in again as fast as he can pull, first passing it
quickly through the bowline on Muliao's rod.
But already a *pala*, a long, slender, scaleless
fish, six times as big as the biggest salmon ever
caught, and with teeth like a rip-saw, has heard
the splash, and is speeding after the decoy.
Deftly the dead fish is drawn through the trap,
followed by the eager jaws and round head and
shoulders of its pursuer. Then, whish ! the
bowline jerks, slips over his smooth, rounded
body, and tightens in a fatal grip upon the
broad, bony tail. And then there is a mighty
struggling, and splashing, and leaping, and the
canoe shoots hither and thither as the crew haul
on the line ; for a full-grown *pala* is as strong
as a porpoise. At last, however, he is dragged
alongside, and then Muliao, grasping a heavy
turtle-spear in his right hand, rises to his feet
and watches. And then, with arm of strength
and eye of hawk, the spear is sped, and crashes
through the *pala's* bony head.

" *Aue !* " and Muliao leans pantingly back in the stern. " Pull him in, my friends, and then let us to the shore. To-morrow, if the day be fair, shall we fish together on Tia Kau, and with God's blessing and the help of the white man's tobacco catch many fish."

The Areois.

A FEW years ago, during the stay of one of Her Majesty's ships at Huahine in the Society Islands, there came on board, to pay his respects to the commander, an old white trader. He was accompanied by an ancient native, who, he said, was his wife's grandfather. The old islander, although nearly bent double with age, was very lively in his conversation, and spoke English with ease and correctness. Captain M——, after discussing the state of the island with the trader, inquired of him where the old native had learned to speak such excellent English. "I suppose," he added, "he was one of the earliest converts to Christianity?"

The trader laughed. "No, indeed, sir. There's not much of the missionary about Matapuupuu. He did gammon to be converted once, but he soon went back to his old gods again. Why, sir, that old fellow was chief priest here once, and the first man of the Areoi society—and he's the only one left now."

Captain M—— had heard of that mysterious body whose power in the early days of missionary effort was so great and so far-reaching in its terrifying and degrading influences as to at one time bring the spread of Christianity to a standstill. He therefore looked at the native before him with unusual interest ; and, as if aware of what was passing in the officer's mind, the object of his scrutiny raised his head and laughed. "Yes, sir, I am the last man of the Areoi on Huahine. There are one or two more of us on Taiarapu, in Tahiti."

"Why have you not become a Christian in your old age, then, now that there are no more Areois left? What good can it do you to remain a heathen?"

Old Matapuupuu shrugged his wrinkled shoulders—"What is the good of Christianity to me now? I am too old to get anything by being a Christian. It is better for me to be an Areoi. I am very old and poor, although I made a lot of money when I was sailing in the whaleships. But, although I am so poor, I get plenty to eat, for the people here are afraid of me. If I became a Christian they would give me nothing to eat, for my power over them would be gone."

" But I should be ashamed to have it known that you belonged to such a wicked lot of scoundrels, old man," said Captain M—— with assumed severity ; " everything that was done by the Areois was bad. Had not their power been broken by the missionaries there would have been no more people left in these islands in another twenty years after they had settled here."

" Bah," answered the old ex-priest, derisively, " that is only missionary talk. There have been Areoi since first men were born. And, see, the people liked us ; for we gave them songs, and music, and dancing. It is true that we made the women who bore us children kill them ; but that was wisely done ; for these islands are but little places, and but for us there would have come a time when the people would have eaten each other for hunger. It is better that useless children should die than grown people should starve."

Half an hour later the trader and the old ex-

priest Areoi bade the captain goodbye, and the
officer, as he watched them going over the side,
turned to the ship's doctor and said with a
laugh, " What an unmitigated old heathen."

" But there's a good deal of sound logic
in his contentions," replied the doctor, seriously.

.   .   .   .   .

The history of the Areois of Polynesia and
the Uritois of the Micronesian Islands is an
interesting subject, and Mr. Ellis in his
" Researches " has given us a full account of
the former ; while Padre Canova, a Jesuit
missionary who was killed in the Caroline
Islands before the time of Cook, has left on
record an account of the dreaded and mysterious
Uritois society of that archipelago.   The Areois
are now extinct, but the Uritois, whose practices
are very similar to those of the Polynesian
fraternity are still in existence, though not
possessed of anything like the power they
wielded in former days.

" The Areois of Polynesia," says Mr. Ellis,
" were a fraternity of strolling players, and
privileged liberties, who spent their days in
travelling from island to island, and from one
district to another, exhibiting their pantomimes,

and spreading a moral contagion throughout society." (Each band or section of the society was called a " mareva," corresponding with the Samoan " malaga "—a party of travellers ; and, indeed, in Australian parlance they might have been designated as larrikin " pushes.") " Before the company set out great preparation was necessary. Numbers of pigs were killed and presented to the god Oro ; large quantities of plantains and bananas, with other fruits, were also offered upon his altars. Several weeks were necessary to complete the preliminary ceremonies. The concluding parts of these consisted in erecting, on board their canoes, two temporary *maraes*, or temples, for the worship of Orotetefa and his brother, the tutelary deities of the society. This was merely a symbol of the presence of the gods ; and consisted principally in a stone for each, from Oro's *marae*, and a few red feathers for each, from the inside of his sacred image. Into these symbols the gods were supposed to enter when the priest pronounced a short ' uba,' or prayer, immediately before the sailing of the fleet. The numbers connected with this fraternity, and the magnitude of some of their expeditions will appear from the fact of Cook's witnessing on one occasion, in Huahine, the departure of seventy canoes filled with Areois. On landing at the place of destination they proceeded to the residence of the king or chief, and presented their ' marotai,' or present : a similar offering was also sent to the temple and to the gods, as an acknowledgment for the preservation they had experienced at sea. If they remained in the neighbourhood preparations were made for their dances and other performances.

" On public occasions, their appearance was, in some respects, such as it is not proper to describe. Their bodies were painted with charcoal, and their faces, especially, stained with the ' mati,' or scarlet dye. Sometimes they wore a girdle of the yellow *ti* leaves, which, in appearance, resembled the feather girdles of the Peruvians or other South American tribes. At other times they wore a vest of ripe yellow plantain leaves, and ornamented their heads with wreaths of the bright yellow and scarlet leaves of the ' hutu,' or ' Barringtonia ' ; but, in general, their appearance was far more repulsive than when they wore these partial coverings."

" Upaupa " was the name of many of their exhibitions. In performing these, they sometimes sat in a circle on the ground, and recited, in concert, a legend or song in honour of the gods, or some distinguished Areoi. The leader of the party stood in the centre, and introduced the recitation with a sort of prologue, when, with a number of fantastic movements and attitudes, those that sat around began their song in a slow and measured tone and voice, which increased as they proceeded, till it became vociferous and unintelligibly rapid. It was also accompanied by movements of the arms and hands, in exact keeping with the tones of the voice, until they were wrought to the highest pitch of excitement. This they continued until, becoming breathless and exhausted, they were obliged to suspend the performance.

Their public entertainments frequently consisted in delivering speeches, accompanied by every variety of gesture and action ; and their representations, on these occasions, assumed something of the histrionic character. The priests and others were fearlessly ridiculed in these performances, in which allusion was ludicrously made to public events. In the "tapiti," or "Oroa," they sometimes engaged in wrestling, but never in boxing ; that would have been considered too degrading for them. Dancing, however, appeared to have been their favourite and most frequent performance. In this they were always led by the manager or chief. Their bodies, blackened with charcoal and stained with " mati," rendered the exhibition of their persons on these occasions most disgusting. They often maintained their dance through the greater part of the night, accompanied by their voices, and the music of the flute and drum. These amusements frequently continued for a number of days and nights successively at the same place. The " upaupa " was then terminated, and they journeyed on to the next district or principal chieftain's abode, where the same train of dances, wrestling, and pantomimic exhibitions was repeated.

Several other gods were supposed to preside over the " upaupa " as well as the two brothers who were the guardian deities of the Areois. The gods of these diversions, according to the ideas of the people, were monsters in vice, and, of course, patronised every evil practice perpetrated during such seasons of public festivity.

Substantial, spacious, and sometimes highly ornamental houses were erected in several districts throughout the islands, principally for their accommodation and the exhibition of the Areoi performances. Sometimes they performed in their canoes as they approached the shore; especially if they had the king of the island or any principal chief on board their fleet. When one of these companies thus advanced towards the land, with their streamers floating in the wind, their drums and pipes sounding, and the Areois, attended by their chief, who acted as their prompter, appeared on a stage erected for the purpose, with their wild distortions of persons, antic gestures, painted bodies, and vociferated songs, mingling with the sound of the drum and the flute, the dashing of the sea, and the rolling and breaking of the surf on the adjacent reef, the whole must have presented a ludicrous but yet imposing spectacle, accompanied with a confusion of sight and sound, of which it is not easy to form an adequate idea.

"The above were the principal occupations of the Areois; and in the constant repetition of these often obscene exhibitions they passed their lives, strolling from the habitation of one chief to that of another, or sailing among the different islands of the group. The farmers, *i.e.*, those who owned plantations, did not, in general, much respect them " (but they feared them), " but the chiefs, and those addicted to pleasure, held them in high estimation, furnishing them with liberal entertainments, and sparing no property to gratify them. This often proved the cause of most unjust and cruel oppression to the poor cultivators. When a party of Areois appeared in a district, in order to provide daily sumptuous entertainment for them, the local chief would send his servants to the best plantations in the neighbourhood, and these grounds, without any ceremony, they plundered of whatever was fit for use. Such lawless acts of robbery were repeated every day, so long as the Areois continued in the district ; and when they departed the gardens exhibited a scene of desolation and ruin that, but for the influence of the chiefs, would have brought fearful vengeance upon those who had occasioned it.

A number of distinct classes prevailed among the Areois, each of which was distinguished by the kind or situation of the tatooing on their bodies. The first or highest class was called "Avae parai," painted leg; the leg being completely blackened from the foot to the knee. The second class was called "Otiore," both arms being marked from the fingers to the shoulders. The third class was "Harotea," both sides of the body, from the armpits downwards, being tattooed. The fourth class, called "Hua," had only two or three small figures, impressed with the same material, on each shoulder. The fifth class, called "Atoro," had one small stripe tattooed on the left side. Every individual in the sixth class, called "Ohemara," had a small circle marked round each ankle. The seventh class, or "Poo," which included all who were in the noviciate, was usually denominated the "Poo faarearea," or pleasure-making class, and by them the most laborious part of the pantomimes, dances, etc., was performed; the principal or higher order of Areois, though plastered over with charcoal, were generally careful not to exhaust themselves by physical effort for the amusement of others.

Like the society of the Uritoi (the Uritoy of the Jesuit Canova), the Areoi classes were

attended by a troop of what may be termed
camp-followers, who, as Ellis observes, " attached themselves to the dissipated and wandering fraternity, prepared their food and their
dresses, and attended them on their journeys
for the purpose of witnessing their dances and
sharing in their banquets. These people were
called Fanaunau (*i.e.*, propagators), because they
did not destroy their offspring, which was in-
dispensable with the regular members of the
whole seven classes." Curiously enough, while
steeped in every imaginable wickedness, there
was with the Areoi a rigid code of morality
among themselves. "Each Areoi, although
addicted to every kind of licentiousness, brought
with him his wife, who was also a member of
the society. And so jealous were they in this
respect, that improper conduct towards the
wife of one of their own number was sometimes punished with death." At Tahaa, in the
Society Islands, a young girl, wife of one of
the class called " Harotea," who had miscon-
ducted herself with a lad at Fare, in Huahine,
was taken before the assembled band of Areois,
and deliberately slain by the leader, who was
her uncle. Her husband, who begged for her

life, met the same fate, as an unworthy member
of the society. "Singular as it may appear,
the Areoi institution was held in the greatest
repute by the chiefs and higher classes ; and,
monsters of iniquity as they were, the grand-
masters, or members of the first order (the
' Avae parai ') were regarded as a sort of super-
natural beings, and treated with a corresponding
degree of veneration by many of the vulgar and
ignorant. The fraternity was not confined to
any particular rank or grade in society, but was
composed of individuals from every class of
people. But although thus accessible to all,
the admission was attended with a variety of
ceremonies ; a protracted noviciate followed ;
and it was only by progressive advancement
that any were admitted to the superior dis-
tinctions.

"It was imagined that those "—to continue
Ellis—"who became Areois were generally
prompted or inspired (by their tutelar gods)
to adopt this course of life. When, therefore,
any individual wished to be admitted to the
ranks of the Areois, he repaired to some public
exhibition in a state of apparent _neneva_ or
derangement. Round his or her waist was a

girdle of yellow plantains or *ti* leaves; his face
was stained with *mati*, or scarlet dye ; his brow
decorated with a shade of curiously painted
yellow coconut leaves ; his hair perfumed with
powerfully scented coconut oil, and ornamented
with a profusion of fragrant flowers. Thus
arrayed, disfigured and yet adorned, he rushed
through the crowd assembled round the house
in which the actors or dancers were performing,
and, leaping into the circle, joined with seeming
frantic wildness in the dance or pantomime.
He continued thus in the midst of the per-
formers until the exhibition closed. This was
considered an indication of his desire to join
their company ; and, if approved, he was ap-
pointed to wait, as a servant, on the principal
Areois. After a considerable trial of his natural
disposition, docility, and devotedness in this
occupation, if he persevered in his determination to join himself with them, he was
inaugurated with all the attendant rites and
observances.

" This ceremony took place at some *taupiti*,
or other great meeting of the body, when the
principal Areoi brought forth the candidate
arrayed in the *ahu haio*, a curiously-stained

sort of native cloth, the badge of their order, and presented him to the members, who were convened in full assembly. The Areois, as such, had distinct names, and, at his introduction, the candidate received from the chief of the body the name by which in future he would be known among them. He was now directed in the first instance to murder his children—a deed of horrid barbarity—which he was in general only too ready to perpetrate. He was then instructed to bend his left arm, and strike his right hand upon the bend of the left elbow, which at the same time he struck against his side, whilst he repeated the song, or invocation, for the occasion. He was then commanded to seize the waist-cloth worn by the chief woman present, and by this act he completed his initiation, and became a member of the seventh, or lowest class.

" There can be no doubt that the desire of females to become members of this strange association was caused by the many privileges it afforded them. The principal of these was that, by becoming an Areoi, a woman was enabled to eat the same food as the men ; for the restrictions of the *tabu* upon women in this

respect were very severe. Females, even of
the highest rank, were prohibited, on pain of
death, from eating the flesh of animals offered
to the gods, which was always reserved for the
men ; but once admitted to the ranks of the
Areois, they were regarded as the equals of
men in every respect, and partook of the same
food."

And so these people travelled about from
village to village, and from island to island, and
sang and danced, and acted for days together ;
but though these "were the general amusements
of the Areois, they were not the only purposes
for which they assembled." They included

"All monstrous, all prodigious things."

The Jesuit Canova, in the account he gives
of the Uritois of the Caroline Islands, says :—
"It is absolutely impossible for the average
human mind to conceive the frightful cruelty,
the hideous debauchery, and unparalleled licentiousness to which these people surrender
themselves when practising their soul-terrifying
rites."

Yet their power and influence were extraordinary. In their journeyings to and fro among the islands they would sometimes locate themselves among a community who were totally unacquainted with them save by hearsay, and who regarded their advent with feelings of terror ; yet, before long, numbers of these same people would desire to, and did enter their ranks. " In their pastimes, in their accompanying crimes, and the often-repeated practices of the most unrelenting, murderous cruelty, these wandering Areois passed their lives, esteemed by the people as a superior order of beings, closely allied to the gods, and deriving from them direct sanction, even for their heartless murders. Free from labour or care, they roved from island to island, supported by the priests and the chiefs ; and often feasted on plunder from the gardens of the industrious husbandman, while his own family was not infrequently deprived thereby for a time of the means of existence. Such was their life of luxurious and licentious indolence and crime. And · such was the character of their delusive system of superstition that for them too was reserved the Elysium which their fabulous mythology taught them to believe was provided in a future state of existence for those so pre-eminently favoured by the gods."

That such a deadly and satanic delusion should be implanted and fostered in the minds of a naturally amiable, hospitable, and intelligent race can only be accounted for by the belief of the sacred source from which it sprang, *i.e.*, the mandate of Oro ; and its destruction by the advances of civilisation had a profound effect on the minds of those who had witnessed the terrible deeds perpetrated when the Society Islands lay under the terror of the Areois.

Australia's Heritage; The New Hebrides Group.

EVERY now and again the Australian colonies are disturbed by a rumour that the present Anglo-French convention for the " control " (whatever that may mean) of the New Hebrides is about to terminate, and that one of the best and most fertile of the island groups in the South Pacific will be annexed by France, which is hot to possess them. Perhaps the inception of this rumour may be due but to the nearing prospect of Australian Federation, which would necessarily revive in the public mind the fact that a few years ago two French men-of-war landed some hundreds of soldiers, virtually took possession of the whole group, and were only withdrawn on the united protest of the Australian colonies to the Imperial Government. From that time began the present

dual control—*i.e.*, the patrolling of the group
by ships of war of both nations—and a very
unsatisfactory arrangement it has proved. Five
years ago, the late Governor of Fiji, in his
capacity of High Commissioner for the Western
Pacific, when questioned as to the claims of the
British settlers in the New Hebrides, and to the
possibility of the group being annexed by either
Power, said, "I cannot tell how the matter will
be settled. Both France and England want the
New Hebrides ; each nation is determined that
the other shall not get it. In the meantime
things must, of course, go on as they are
doing." And things have been going on very
unsatisfactorily, in the opinion of the men who
have made the group what it is—the English
settlers, traders, planters, and merchants. It is
not my purpose, however, to enter into the
rival claims of the English and French residents,
but to give a brief description of the islands
themselves. Yet one thing may be said, and
that is this : The group was opened up and
surveyed by British ships ; British and Australian money has done a great civilising work
there ; the men who first discovered them to
commerce were Englishmen ; the natives are

ardently desirous that England should annex their islands ; and their occupation by France for the extension of her malign convict system will constitute a menace to the Australian colonies, leaving alone the danger to them which the possession of such a magnificent base as these islands would give to a Power who may some day be at war with England. But now as to the group itself.

Next to the Fijis and the Solomon Islands, the New Hebrides are the finest cluster of islands in the South Pacific ; and were British settlers in the group freed from their present harassing disabilities in the way of employing native labour to work their plantations it would leave Fiji far behind in the development of its abounding resources. It possesses magnificent harbours, forests of timber awaiting the axeman and saw-miller, land suitable for coffee and cotton, and other tropical products, a climate that is no hotter than that of Ceylon or Samoa, and a native population which, within two years after the declaration of British sovereignty, would, owing to the influence of the English missionaries in the group, be as amenable to the precepts of civilisation as the too-highly

Christianised people of Tonga and Fiji, where
to-day a man's life and property are as safe
as if he lived under the shadow of Westminster.

The largest island of the group is Santo—
the Espiritu Santo of Quiros, who in a memoir
to his Royal master, Philip III., spoke of it as
"a very great island : a country of the richest
fertility and beauty. It is to my mind one of
the finest in the world, and capable under
colonisation of becoming one of the richest
places in the Southern Hemisphere." Its length
exceeds eighty miles with an average of thirty-
two in width, and within the great sweep of the
mighty barrier reef that encloses it are some
scores of clusters of low-lying islands of purely
coral formation, densely covered by groves of
coco-palms, and inhabited by a numerous
population of strong athletic savages of
Melanesian blood, whose earliest recollections
of white men date from the old colonial days,
when the group was visited by sandal-wooding
ships from Sydney and vessels engaged in the
colonial whale fishery. The present race that
people the mainland are, no doubt, new comers
within the past three hundred years, for on
several parts of the islands there are traces of

occupation by an earlier race : detached pillars composed of large stones, long stretches of broken walls indicating walled towns, and fragments of rough masonry cemented with a chunam of coral lime and river sand. Much of Santo is covered by noble forests of timber, and the littoral is of remarkable fertility. The inhabitants, though nearly all thorough savages, have a better reputation than most of the people inhabiting the larger islands of the group, and the record of white men who have lost their lives in the group is low on Santo.

An island with an evil reputation in the past is Tanna, for the inhabitants are savage and treacherous, and though there are English traders living among them in security, the Tanna people have perpetrated some fearful massacres upon vessels engaged in trading and recruiting native labour. The whole island is a magnificent panorama of tropical island beauty, thirty-five miles long by eleven in width. Towards the southern end the land trends away in a gradual slope from lofty mountains, densely wooded and enveloped in mist and clouds. The low-lying coast lands are of surpassing fertility, and millions of coco-palms encompass the shores, while about the thickly populated villages are carefully tended plantations of sugar-cane, bananas, pineapples, yams, taro, and other tropical vegetables and fruit. An old New Hebrides trader captain estimates the population of Tanna at 8,000, and believes it could support 30,000.

Mallicolo—a favourite recruiting ground for the Fijian and Queensland vessels employed in the labour trade—is about fifty-five miles in length, with an average width of twenty miles, and is covered with magnificent forests from its littoral to the very summit of the range that traverses the island. That the timber of Mallicolo is valuable for ship-building purposes Australian trading captains well know, but until the group is annexed by either France or England there is but little prospect of its forests being exploited in a systematic manner. The island is intersected by some splendid streams of water ; and although malarial fever is prevalent at certain seasons of the year, the climate on the whole may be considered healthy. It is a favourite rendezvous for both the English and French war-vessels engaged in patrolling the group, and is also frequently visited by trading vessels from Sydney and those in the service of the French New Hebrides Company—an association that while possessing a considerable amount of purchased land in the various islands is merely kept going by a subsidy from the French Government.

Aneityum is a high and mountainous island with but a narrow belt of low-lying land running round its coast, but it possesses some splendid forests. A saw mill, owned by an Australian firm, has been established here and some excellent timber has been produced for local purposes—house-building, ship-building, etc. This island is but twelve miles in length and about six in width. The natives are all (alleged) Christians and number about 2,200. They were the first to come under missionary influence, and are a peaceable, law-abiding race. They are, of course, of Papuan blood, and of a very dark colour, with woolly or frizzy hair like that of most Fijians. No harbours for large vessels are available, but the steamers of the Australasian New Hebrides Company and small sailing craft find anchorage under the north-west end of the island during the strong south-east trade winds.

6

The seismic forces of nature are much in evidence in the New Hebrides group. On Tanna there is a volcano on the south-east end of the island that is frequently in a state of commotion. Viewed from seaward on a dark night it presents a weird and awe-inspiring spectacle. Rumblings, groanings, and dull roaring sounds emanate from its interior, and the noise of its restless convulsions can be heard at Aneityum, nearly fifty miles distant. The volcano itself presents an impressive sight even in daylight, rising as it does to a thousand feet, the grim reddish-brown of its perfect cone affording in its barren sides a startling contrast to the amazing wealth of verdure that, despite its fierce eruptions, prevails everywhere around it. The mighty forces that lie in its heart are seldom quiet; and at short intervals a straight column of smoke, dark, heavy, and pall-like, shoots upward, till, as it ascends, a canopy is formed. This, in the course of half an hour or so, expands and unfolds itself till it resembles a gigantic aërial mushroom. Then it gradually disperses; hollow groanings and deep rumblings follow, and then, as the black sulphurous smoke changes to a pale blue, there again comes a sudden convulsion, and a fresh pillar of inky smoke shoots high in the air.

Erromanga will always be associated with the name of John Williams, the pioneer missionary to the South Seas, who was there murdered with his colleague, Mr. Harris, in 1839. In later years four other missionaries yielded up their lives to the savage inhabitants, the last being Mr. Gordon, who was killed there in 1872. At the present time the Erromangans are, with the exception of the people of the little island of Tongoa, the most thoroughly Christianised inhabitants of the group.

At Vaté, the *entrepôt* of the whole group, is established the principal trading centre of the New Hebrides, and here live the greater portion of the English and French settlers established in the group. Vaté is thirty-five miles long by eight broad, of moderate elevation, has some noble harbours, a fertile soil, splendid banana plantations, and, in a small way, is the Ceylon of the South Seas.

Jack in the Atolls.

HISTORY does repeat itself. The story of the Cornish clergyman who in the middle of his discourse jumped down from his pulpit, and, imploring his hearers to "start fair," raced them to the scene of a promising wreck, has its Polynesian counterpart—clergyman, church, and all. Some little difference there is, however, with regard to other accessories of the South Sea story ; as the coloured minister, instead of the regulation surplice and black trousers, wore a white shirt only, and trousers were a missing quantity. He was, as I have said, a native clergyman, and lived and laboured—"laboured" is merely euphemistic, as any one knows who has knowledge of native teachers—on one of the atolls in the Caroline Islands. Service had commenced, and Miti Paulo Ionatani (*Anglicè*—the Reverend

Paul Jonathan) had just given out the first hymn, when there was a sudden commotion among his squatting congregation. A native, his bronzed skin streaming with perspiration and his frame panting with excitement, had put his head and shoulders through one of the low, wide windows of the sacred edifice (from the outside, of course), and the Reverend Paul, in severe but dignified tones, called him an unmannerly pig, and then asked him what he wanted.

"The sharks are coming in, your reverence!"

In an instant the deep religious calm of the congregation was broken up, and half a minute later the church was cleared in a mad rush to get to the beach, launch the canoes, and go a-fishing for sharks, the minister following as hard as he could run, divesting himself of his garment of office by the way. Like his Cornish prototype, he meant to have a share of the plunder. (I wonder whether the Cornish story originated from the Polynesian story, or *vice versâ.* Both are true.)

But shark-catching means money down there in the Carolines and the equatorial atolls of the North and South Pacific; and sometimes vast numbers of sharks, swimming together in "schools," like sardines, enter the lagoons at certain seasons of the year and cause no end of excitement among the brown-skinned people ; just as much, in fact, as that which occurs when a "school" of bottle-nosed whales is driven ashore by the inhabitants of the Faröe Islands.

Every now and then one may see noted in Australian papers the arrival of an island trading vessel bringing, among other cargo, so many tons of shark-fins ; and the uninitiated naturally wonder for what on earth shark-fins are brought to the marts of civilisation. That is easily answered—they are regarded as a great delicacy by John Chinaman. (By the way, it seems an oversight that no one in England thought of presenting Li Hung Chang, when he visited England a year ago, with a string of shark-fins in return for his inexhaustible presents to the British aristocracy of packets of tea ; a dozen or so—especially if not quite dried—would have moved him greatly.)

For the last fifty years shark-catching has been followed on a large or small scale by the inhabitants of the South Sea Islands, from Tonga in the south to the beauteous Pelews in the far north-west. Until of late years only the fins and tails were cut off, dried on strings, and sold by the natives to either resident traders or wandering trading vessels. By these latter they are taken to Sydney, and there sold to Chinese merchants, who in their turn ship them home to China. But nowadays not only are the fins and tails dried by the natives in increasing quantities, but the whole skin is stripped off, pegged out like a bullock's hide, and sold to the white men. But the skins do not go to China. They are sold to German trading vessels, and no one even to this day knoweth for what purpose they are used; some new process of tanning the intractable cuticle of Jack Shark has been discovered in Germany, it is said. No one knows more than this; probably the only man who *does* know is that modern Lokman the Wise, the Emperor William: may he tell us dull English people all about it some day when he, in his Improvement-of-the-Universe Scheme, writes us something on the subject of cross-breeding in sharks, whereby a toothless and amiable variety may replace the present breed, which have no manners to speak of and are always hungry.

But I want to say something of how and where sharks are caught and of those who catch them.

In the high, fertile islands of the North and South Pacific, such as Samoa, the Hervey Group, and the Society Islands there is but little of this dangerous fishing done. Nature there is too bounteous to the brown-skinned people. Born to a fruitful soil, with abundance of both vegetable and animal food, the natives have no need to exploit the ocean day and night in order to live, as do the wild, sun-baked denizens of the low-lying Equatorial atolls of the Gilbert and Marshall groups and the countless coral islets of the Western Carolines, where the people know naught of the joys of the mealy yam or taro, and the toothsome baked bread-fruit and sucking-pig are not. For there is nothing to eat on such islands as those but coconuts and fish, varied occasionally by *puraka*—a huge, coarse vegetable as thick as an elephant's leg, with a touch of *elephantiasis* thrown in.

But there are plenty of sharks. They swarm. Go out in a canoe at night-time, anywhere in one of the lagoons, light a torch of *au lama* (dried coconut leaves), and look.

Perhaps you may only see one or two at first, swimming to and fro at a few fathoms' depth; in ten minutes you may see fifty! and they are all hungry. A bad short time would a man have did he fall overboard at night. In daylight the natives know no fear of Jack, but they do not like getting capsized in the darkness; and the darker the night the more danger. And even when he is young, and not a fathom long from his nose to his tail, Jack can snap off the arm of a full-grown man as easily as a man can swallow an oyster.

So, there being plenty of sharks, the Ellice, Gilbert, or Marshall islander is resigned to the poverty of his island soil, catches his shark, and is thankful. For he sells Jack's fins and tail to the trader for tobacco, calico, guns, ammunition, and gin—when gin can be bought; and his wife, when she meets her brown-skinned lord and master on the beach as he returns from fishing, looks anxiously into the blood-stained canoe to see how many *kapakau* (fins) he has taken. Two or three dozen or so, when dried, may mean that lovely hat trimmed with violent green ribbon on a bilious red and yellow ground that the trader showed her one day. Then she picks up the "take," puts it into a basket, and an hour later Jack's motive power is suspended on a cinnet line between two coconut trees, drying for market.

All the people of the Gilbert Islands are expert shark fishermen ; but the men of Paanopa (Ocean Island) claim to be, and are, *facile princeps* in the forcible art of clubbing a shark before he knows what is the matter with him, and what the horrid thing is that has got into his mouth.

First of all, though, something about Ocean Island itself. It is but a tiny spot, rising abruptly from the sea, about 300 feet in height, situated fifty miles south of the Equator, and in 168 deg. 25 min. east longitude, and inhabited by a fierce, turbulent race of dark-skinned Malayo - Polynesians, allied in want of manners and fulness of beastly customs to their Gilbert Island neighbours, three hundred miles to the windward. Half a cable's length from the land itself, and not twenty yards from the flat shelving coral reef that juts abruptly out from the narrow strip of beach, the water is of great depth—fifty, in some places ninety, fathoms deep.

At the first break of dawn the men, naked save for a girdle of grass around their loins, sally out from their grey-roofed houses of thatch, and launch their canoes for the day's work. Wonderful canoes these are, too—mere shells composed of small strips of wood sewn together with coconut cinnet. In no one of them will you see a plank more than two feet in length and six inches in width ; many are constructed of such small pieces of wood so deftly fitted and sewn together that one wonders how the builders ever had the patience to complete the craft. But wood is scarce on Ocean Island ; and whenever—as sometimes happens —a canoe is smashed by the struggles of a more than usually powerful shark, the tiny timbers are carefully picked up by other canoes and restored to the owners, who fit them together by degrees until a new hull is pieced together.

Perhaps twenty or more canoes go out together. No need to go far. Just outside the ledge of the reef is enough, for there Jack is waiting, accompanied by all-sized relatives, male and female. Lying upon the little grating of crossed sticks that reaches from the outrigger to the gunwale is the tackle. Rude it is, but effectual—a huge wooden hook, cunningly trained when it was a young tree-root into growing into the proper shape, and about forty fathoms of strong coconut-fibre rope—as thick as whale-line and as strong. Taking a flying fish, or a piece of the flesh of a shark caught the previous day, a native ties the bait around the curve of the great hook. Then he lowers the line, which sinks quickly enough, for the wooden hook is as heavy as it is big. Presently the line tautens— Jack is there. The steersman strikes his paddle into the water to bring the canoe's head round, the man holding the line gives it a sudden jerk that makes the outrigger rise a foot out of the water and nearly upsets the little craft, and a third native handles a short iron-wood club expectantly. Perhaps, if Jack is a big fellow, he will obstinately refuse to turn, and make a strenuous effort to get away deep down into the blue gloom, a hundred fathoms below. Sometimes he does ; apparently nothing short of a steam-winch at the other end of the line would then stop him ; and so fathom by fathom the line descends, and the steersman and "clubber" look anxiously at the few fathoms left coiled up on the outrigger platform. Generally, however, Jack is turned from his direct downward course by a sudden jerk. Then all hands " tail on " to the line to get him to the surface before he gets his head free again for an attempt at another dive.

Meanwhile, every other canoe has got fast to a shark, and now there arises wild clamour and much bad language as the lines get foul, and canoes bang and thump against each other. Perhaps four or five will be in a lump, together with one or two sharks lashing the water into foam in the centre and turning over and over with lightning-like rapidity in the hope of. parting the line or smashing the outrigger. This latter is not a nice thing to happen, and so the clubmen anxiously watch for a chance to deal each struggling brute a blow on the head. Often this is not easily effected, and often too it is not needed, for the shark may let his tail come within the reach of the steersman's arm, and a slashing blow from a heavy-backed, keen knife takes all the fight out of Jack—at one end, at any rate ; if it is only a young fish, however, the tail is grasped by a native and cut off before Jack knows that he has lost it.

By and by those natives who are fast to a
big fellow call out to their comrades that their
shark is too heavy and strong to bring alongside and kill, and ask for an implement known
to whalers as a "drogue"—a square piece of
wood with a hole through the centre which,
attached to the end of a line, gives such
resisting power that the shark or whale
dragging it behind him is soon exhausted. So
the "drogue" is passed along from another
canoe, and being made fast to the end of a small
but strong line, the canoe is carefully hauled up
as near as possible to savage, struggling Jack.
At the loose end of the line is a noose, and
watching a favourable moment as Jack lifts his
tail out of the water, the steersman slips it over,
and away goes line and "drogue"—the man
who is holding on to the main line casting it all
overboard so as to give the shark plenty of room
to exhaust himself. In ten minutes more he is
resigned to his fate, gives in, is clubbed in peace
and towed ashore—that is, if his ocean prowling
friends and relatives do not assimilate him unto
themselves before his carcase is dragged up on
to the reef, and skinned by the savage-eyed
Ocean Island women.

The Cutting off of the "Boyd."

IN the *Sydney Gazette* of August, 1809, there appears a notice of the arrival of the ship *Boyd* with a cargo of convicts for New South Wales. She had sailed from the Thames on March 10, 1809, and arrived in Sydney Cove on August 14th following. After refitting she left in November on her return voyage to England, but proceeded viâ New Zealand, having been chartered by Mr. S. Lord, of Sydney, to touch at the port of Whangaroa and load a cargo of kauri spars for the naval authorities at the Cape of Good Hope. Mr. Lord also put on board a large quantity of New South Wales mahogany, sealskins, oil, and coal for the same market, in all amounting to the value of £15,000. There was among the other passengers "an East Indian captain named

Burnsides, who having by industry accumulated a fortune of £30,000, was on his return to end his days among his friends on the banks of the Liffey."

For those days the *Boyd* was a large ship, well found and well armed, was owned by Mr. George Brown, a London merchant, and was under the command of Mr. John Thompson. In addition to her crew, she carried a large number of passengers, bound to the Cape of Good Hope, and also some convict labourers, who were to be employed in cutting and loading the spars. Her European complement consisted of some seventy persons, and some five or six Maoris, and these latter, it was commonly asserted, and is believed to the present day, were the instigators of the crime, although that designation of the massacre is still resented by the Maoris of the present day, who insist that the slaughter of the *Boyd* people was but a natural and just retribution for the cruel flogging of a chief named Tara by the Europeans.

The first particulars of the event reached Sydney through Mr. Alexander Berry, super-cargo of the ship *City of Edinburgh*, but his testimony, coming entirely from native sources, was incorrect and imperfect in many details. His narrative appeared in the *Sydney Gazette* some time in 1810, and also in the *Edinburgh Miscellany*. Abridged it is as follows :

These are to certify that during our stay in this harbour (Bay of Islands) we had frequent reports of a ship being taken by the natives in the neighbouring harbour of Whangaroa, and that the ship's crew were killed and eaten. . . . Mr. Berry, in order to ascertain the truth of this report, accompanied by Mr. Russel and Metangatanga, a Maori chief, set out for Whangaroa in three armed boats on Sunday, December 31, 1809, and upon their arrival they found the miserable remains of the ship *Boyd*, Captain John Thompson, which the Maoris, after stripping of everything of value, had burnt to the water's edge. They were able to rescue a boy, a woman, and two children, the only survivors of this shocking event, which, according to the most satisfactory information, was perpetrated entirely under the direction of that rascal Te Pahi (a Maori chief who had been made much of by the then Governor of New South Wales).

This unfortunate vessel, intending to load with spars, was taken three days after her arrival. On the day of her arrival she was boarded by a great number of Mowrees (*sic*), who expressed their eagerness to assist Captain Thompson and his crew in cutting the spars. Later on in the day Te Pahi arrived from Te Puna, and came on board. He stayed only a few minutes, but in that time took a keen survey of the ship, and noted the loose discipline maintained on board. He then quietly des-

cended into his canoe, but remained alongside the vessel
and gave some quiet instructions to the great number of
natives who surrounded the vessel on all sides.   One by
one these canoes drew up alongside, and their crews, under
the pretext of trading, gradually managed to get on board,
and sit down upon the deck.   After breakfast Captain
Thompson ordered two boats to get ready, and left the
ship under the guidance of a native to look for spars.
Presently Te Pahi, with the utmost calmness, ascended to
the deck and surveyed the scene.   Satisfied that the proper
moment had arrived for his bloody purpose, he gave the
signal.   In an instant the apparently peaceful natives, who
had been sitting quietly upon the decks, rushed upon the
unarmed crew, who were dispersed about the ship at their
various employments.   The greater part were massacred
in a moment, and were no sooner knocked down by blows
from meres (clubs) than they were cut up while still alive.
Five or six of the crew escaped into the rigging.   Te Pahi
now having possession of the *Boyd*, hailed these with a
speaking trumpet, and ordered them to unbend the sails
and cut away the rigging and they should not be hurt.
They complied with his commands, and then came down.
He then took them ashore, and afterwards killed them,
The master, who had gone ashore unarmed, was easily
despatched.

The names of the survivors are : Mrs. Nanny Morley
and child, Betsy Broughton, and Thomas Davis (boy).
The natives of the spar district of this harbour (Kororareka,
Bay of Islands), have behaved well, even beyond expecta-
tion, and seem much concerned on account of the unfortunate event ; and dreading the displeasure of King

George, have requested certificates or their good conduct in order to exempt them from his vengeance. But let no man after this trust a New Zealander.

Simeon Pateson, Master.

Alex Berry, Supercargo.

James Russell, Mate.

Ship *City of Edinburgh*, Bay of Islands, January 6, 1810.

Not long after this Captain Chase, of the *Governor Bligh*, was able to obtain further particulars from a native of Tahiti, who was one of the *Boyd's* crew, and had probably been spared on account of his colour. According to this man's account, which appears to be authentic, the captain was accompanied by the chief mate, and three, not two, boats were manned to get the spars on board. Among those who were with the landing party were the six Maori seamen from the *Boyd*. These were the men whom it is alleged that Captain Thompson ill-treated. (Nothing definite to this effect seems to have been proved.) The boats were conducted to a river, on entering which they were out of sight of the ship ; and after proceeding some distance up, Captain Thompson was invited to land and mark the trees he wanted. The boats landed accordingly, the

tide being then beginning to ebb, and the crews
followed to assist in the work. The guides led
the party through various paths that were least
likely to answer the desired end, thus delaying
the premeditated attack until the boats should
be left high and dry, and the unfortunate crews
unable to escape. This part of their horrible
plan accomplished, the natives became very
insolent and rude, ironically pointing at decayed
fragments of trees and inquiring of Captain
Thompson whether such would suit his purpose
or not. And then the Maori seamen of the
Boyd threw off the mask, and in opprobrious
terms upbraided him with their maltreatment on
board, informing him at the same time that he
should have no spars there but what he could cut
and carry away himself. Captain Thompson,
apparently caring nothing for the disappointment, turned carelessly away with his people
towards the boats, and at this moment they
were savagely assaulted with clubs and axes,
which the assailants had until then concealed
under their dresses, and, although the boats'
crews had several muskets, yet so impetuous
was the attack that every man was slaughtered
before one could be used.

This dreadful deed accomplished, the bodies of Captain Thompson and his unfortunate men were cut up, cooked, and devoured by the murderers, who, clothing themselves with their victims' apparel, launched the boats and proceeded towards the ship, which they determined to attack. It being very dark before they reached her, and no suspicion being entertained of what had happened, the second officer hailed the boats, and was answered by the villains who had occasioned the disaster, that the captain having chosen to remain with his men on shore that night for the purpose of viewing the country, had ordered them to take on board such spars as had already been procured.

Satisfied with this, the second mate ordered the boats to come alongside, and stood in the gangway as the first of the natives ascended the ship's side. In an instant his brains were dashed out with a *mere*, or jade club, and then all the seamen of the watch on deck were in like manner surprised and murdered. Some of the assassins then went down to the cabin door, and asked the passengers and others to go on deck and look at the boat load of spars. An unfortunate female passenger was the first to open her door and come out, and the moment she ascended the companion she was cut down. The noise occasioned by her fall alarmed the people that were in bed, who, running on deck in disorder, were all killed as they went up the companion, except four or five who ran up the rigging and remained there till daylight.

The next morning Te Pahi appeared along-side in a canoe, and was much offended at what had happened, but was not permitted to inter-fere or remain near the ship. The unfortunate men in the rigging called to him and implored his protection, of which he assured them if they could make their way to his canoe. This they at last succeeded in doing, and, although threatened by the other Maoris, the old man landed the white men on the nearest point. But the moment they reached the shore they were surrounded, and Te Pahi was forcibly held while the murder of the unhappy fugitives was perpetrated.

Te Pahi, however, who was a chief of great renown, was afterwards permitted by the people of Whangaroa to take three boatloads of any property he chose out of the ship, firearms and gunpowder excepted, and the bulk they divided among themselves. The salt provisions, flour, and spirits they threw overboard, considering them useless. The muskets they prized very much, and one of the savages, in his eagerness to try one, stove in the head of a barrel of gun-powder, and filling the pan of the piece snapped it directly over the cask ! In an instant the powder exploded, and killed five native women and nine men, and set the ship on fire. Her cables were then cut, and she was allowed to drift ashore.

Of the rescued survivors of this dreadful massacre, Mrs. Morley died ten months later at Valparaiso ; her child, when grown to woman-hood, kept a school in Sydney. Miss Broughton became Mrs. Charles Throsby, of Sydney ; and Davis, the boy, was drowned at Shoalhaven in the colony of New South Wales, in May, 1822. And to-day, when the tide is low, the brown-skinned descendants of the Maoris who cut off the ill-fated *Boyd* will show you her weed-covered timbers protruding from the mud— silent witnesses of one of the many tragedies of the Southern Seas.

My Native Servants.

I HAD just said good-bye to the captain of the trading vessel that had landed me on Niué (Savage Island), and was watching her getting under way, when I saw that a deputation of the leading villagers was awaiting me at the top of the rugged path that led to my dwelling. Seeing among them several of the people who had been assisting the ship's crew to carry my goods up to the house, I concluded they were waiting to tell me the usual native story of having individually and collectively strained their backs by lifting cases of tinned beef, and demand another dollar each.

I was mistaken. Soseni, the spokesman, stepped out and said that he and the deputation, representing as they did the public voice of Avatele village, respectfully desired to warn me against engaging any strangers from Alofi as domestics. They did not want to damage any one's reputation, but the Alofi folk were a bad lot. Certainly there were *some* who were honest —people who originally belonged to Avatele, and——

I said I would not decide just then. I should wait till I had settled down a bit. At present, however, I added, old Lupo's son, Moemoe, would cook for me.

The deputation seemed annoyed, and Soseni said they were sorry for me. Lupo was a very good man, and although a Samoan, was an esteemed fellow-townsman and a deacon as well ; also he had married a Niué woman, which was in his favour. But his sons were two notoriously improper young men (applause) ; their mother, while of spotless morals, was a confirmed cadger and a public wrangler and shrew. As for the daughters—well, I could look at *their* record any time in the *fakafiili's* (judge's) charge-book.

I thanked the deputation for their goodwill, and said that while I would not decide hastily in reference to other servants, I quite intended keeping Moemoe as my cook. I had, I said,

known old Lupo for many years, and long ago
promised him that if ever I came to Savage
Island I would take some of his family into
my service.

"Thank you very much," said the polite
Soseni, speaking in English, and then he and
his friends bade me good-day, and somewhat
hurriedly left me. (I learnt half an hour afterwards that they at once called an emergency
meeting of the town council, and passed a law
that no member of Lupo's family but Moemoe
was to take service with the new white man
under any excuse ; also that if any of Moemoe's
relatives were seen hanging about the premises
on the look-out for a job the general public
were at liberty to stone them.)

Just as I reached the gate leading to the
house I heard angry voices from the back ; then
followed the sound of blows, accompanied by
much bad language, and presently three men
and four women rushed down the path, pursued
by a hundred or so of people of both sexes,
who assailed the fugitives with showers of
stones. Old Lupo came out to meet me.

"What is all this row, Lupo?" Lupo
smiled pleasantly and said it was nothing—

"only some man and womans, sir, from Alofi.
They wanted to come inside and talk to you
about getting some servants from their town.
And this made the Avatele people cross ; yes,
sir, very cross. So they threw some stone at
them." (I must mention that Lupo always
spoke English to white men, and to address
him in the native tongue was a sore affront.)

"Oh, I see. Well, I'm very hungry ; is my
supper ready ? And, I say, Lupo, don't let any
more people in to-day to talk about servants."

"All right, sir," he replied somewhat uneasily.

I heaved a contented sigh as I mounted the
verandah steps, for the day had been one of
toil,—and I was eager to rest a little before
supper. My little daughter was already asleep
in a fellow trader's house near by, worn out
with the excitement of her novel surroundings.
I stepped into the big sitting-room of my new
abode, and there, sitting on the floor in solemn
silence with their backs to the wall, were about
fifty women. They all smiled pleasantly at me
as I entered, and then all began to talk at once :
each one wanted to be nurse to the *tama-fafine-
toatsi* (little girl).

"Here, I say, Lupo, clear all these women

out of this.   What do they want swarming into
the place like this ?   Tell them that I won't talk
about servants till to-morrow morning.   And,
besides that, I want to eat my supper, and I can't
eat it with fifty women staring at me in a circle."

Old Lupo stroked his bald brown head and
coughed apologetically.   " If you please, sir,
these Niué womans all very much want to
know *who* you going to get for nurse-woman
for your little girl."

On my refusing, with some warmth, to dis-
cuss the matter with them at that moment, and
requesting them to clear out, they unanimously
stated that I was the worst and most ill-
mannered white man they had ever seen ;
furthermore, that if I engaged a single servant,
male or female, from any other village than
Avatele, the blood of that person would be on
my head.   However, they would, they said,
come again in the morning *with some friends*
and talk the matter over with me.

.   .   .   ..   .

A few weeks later, my trading colleague, who
lived ten miles away on the other side of the
island, came along to see how I was getting on.
He had heard that there was some ill-feeling

among the various villages about my servants, and thought he might advise me, as I was a stranger to Niué people. I fell upon his neck and wept down his back, and told him that my servants had got possession of me ; I seemed to have engaged the whole village.

.   .   .   .   .

That confounded young ruffian Moemoe was the worst. He was a tall, well-built youth of nineteen, with a pale olive complexion and big, dreamy eyes that looked soulfully out from the black glossy curls which fell about his forehead. He turned up on the first day dressed in a white duck suit and canvas shoes, and with scarlet hibiscus flowers stuck through his curls, one over each ear. He seemed a clean, intelligent lad, but a bit languid, and said he would be content with five dols. per month ; also that he could make bread. I at once took him over to the detached kitchen, unlocked the door, showed him that all the necessary utensils used by my predecessor were there and in good order, and told him to come to me for kitchen stores. He said, " All right," sat down on a stool and, asking me for my tobacco pouch, began to fill his pipe. Thinking that he probably wanted to meditate a while on the responsibilities of his position, I withdrew.

Hakala, the senior nurse, had been strongly recommended to me by Captain Packenham as a most excellent and deserving woman ; and, more than that, as the widow of a white man who had been hanged in Queensland. She was a pleasant-looking, smiling-eyed woman of about forty, with her long hair dressed *à la Suisse* ; and although she could not speak a word of English, I felt sure my little girl would like her. Besides that she was a widow, and who can resist the claims of the widow upon our pity? I could not. And presently my daughter ran out to her and put out her arms to be lifted up. The woman's eyes, sparkled and danced with pleasure, her brown cheeks dimpled, and a soft, cooing, mother-like laugh gently shook her ample bosom, and then subsided into an endearing, whispering *tuk, tuk, tuk* —just the sweet, crooing sound a mother hen makes as her chicks cuddle up beneath her loving wing. And because of this, and of her widowhood, I gave Hakala the billet of boss nurse. (She entered on her duties at once, and when night came she lay down upon her couch

of mats just under my youngster's bed. About
midnight I looked in and saw that two native
children lay one on each side of her. I
awakened her, and asked her what she meant,
bringing native children in the bedroom.

" They are mine," she said with a smile.

" Why," I said, angrily, " Captain Packenham
said you had no children."

"Only native children, sir. I be married
again now. My husband comes here to-morrow
to live with me. He is a good man and says
he will help Moemoe to make bread for you.")

The next morning I engaged all the servants
I wanted, but had a lot of opposition from the
town councillors because I selected a washerwoman from Alofi, a rough carpenter from
another coast village called Fatiau, and a second
nurse, or rather nursemaid, from a bush town
called Hakupu. She was quite a young thing,
and promised faithfully never to let the little
white girl out of her sight for a moment during
her walks. Her name was E'eu. But, so as
not to cause too much ill-feeling and jealousy
among the Avatele people by the inclusion of
too many strangers, the whole of the labourers,
male and female, that it was necessary for me to

employ about the station were residents of the village. This seemed to satisfy the authorities and all promised well.

At noon I went out to the cookhouse to see how my *chef* was getting on. He had taken off his coat and shirt, but was still sitting down, playing an accordion to an audience of a dozen young women, all more or less in a state of *déshabille*—even for Niué women. They fled wildly the moment I appeared.

"Moemoe, who are those girls? Why did you let them come in here?"

Oh, they were cousins of his, he said, and had come to see him make bread. And he wanted to begin work at once. And would I mind if some of the girls helped him in the kitchen? Minea was good at cleaning knives, Toria wanted to mend a hole in the floor mat, Kahé said she would like to help him peel the yams and taro, and Talamaheke—the girl with the wreath of orange blossoms—wanted to wash up the plates ; the others were willing to make themselves generally useful. Here he was interrupted by a face, apparently his double, appearing at the kitchen window, and the angry exclamation of "Liar!"

"Do not heed *him*," said my *chef*, composedly ; "that my brother, and he very jealous of me. He thinks *he* can make bread."

I warned the intruding brother off the premises, and was just halfway across the grassy sward that separated the kitchen from the dwelling-house, when I heard loud feminine yells, and cries for me to come quickly. Rushing into the big sitting-room a pretty scene was revealed. Hakala, the head nurse, valiantly assisted by the pretty E'eu, was engaged in deadly combat with two other women who were apparently endeavouring to tear her hair out by the roots. My infant daughter was standing on the table, her shrieks of terror only seeming to nerve the combatants to greater efforts to destroy one another. Seizing a canoe paddle from a fat, burly native who stood at the door applauding the struggle, I belaboured the bare legs of the intruders with such effect that both women dropped upon the matted floor and contented themselves with hurling opprobrious epithets at Hakala, and promising to come again later on. In a few moments the house was filled with natives, and an animated discussion took place as to who were the

aggressors. Then Soseni came to my assistance, and, by banging right and left, he with a
heavy stick and I with my paddle, we managed
to clear the room. Then we learned that the
fight arose in a very simple manner. Hakala
had been giving her charge something to eat,
when she was perceived by the two women
from outside the station fence, who told her
she was not fit to take charge of a pig, let
alone a child—and that a white child. This
she very properly resented. .

During the day several minor fights arose
out of trifling matters. The native teacher,
accompanied by his daughter, a huge mass
of adipose tissue, named Pépé (The Butterfly),
solicited an interview with me. The reverend
gentleman said he did not want to harrow
my feelings, but — well, he would let his
daughter speak. And she spoke. She said
that I ought to know that the girl E'eu, who
had part charge of my " beautiful, sweet little
bird," was a sinner of the worst description.
Did I know that she (E'eu) had been turned out
of Sunday School for dancing heathen dances
with some other girls one moonlight night?
I said that did not matter to me. She said that

it would not be good for my child to be taught such things. She herself was a proper girl, and hated wicked and immodest people, like this E'eu. At this my native carpenter, who was working near by, mending a window, laughed derisively, and Pépé's papa asked him what he meant. He replied by making the assertion that Pépé was the giddiest girl in Niué.

"How dare you say that, you pig!" demanded the minister; and then, turning to me, "This man is a very evil-hearted person. He it was who stole the handkerchief of Commodore Goodenough ten years ago." And then the graceful E'eu appeared at the doorway, carrying my infant. In an instant she placed the child in the carpenter's arms, and then flew at the monstrous Pépé like a tiger-cat. We—the teacher and myself—managed to separate them after they had bitten each other savagely.

Later on in the afternoon the washerwoman from Alofi came to me to have her head dressed with sticking-plaster: an Avatele woman had struck her on the temple with a stone. After this matters settled down a bit.

.   .   .   .   .

Two weeks later, E'eu, big-eyed, red-lipped,

and lithe-limbed, asked me if she could take my little girl for a walk to a village called Tamakautoga. "There be people of mine own blood there, master. And they hunger to see this little child of thine. And I shall be careful that she eateth none of those things of which thou hast warned me—pork, or cold taro, or baked *fa'e* (octopus), or guavas, or raw fish."

I saw her lead my little one away under the long, palm-shaded path that led to Tamakautoga, and, having nothing to do in particular, went to the house of a trader near by for a smoke and a chat. From his verandah we commanded a view of the line of reef that stood out like a shelf from the precipitous shore of the island. The tide was half-ebbed, but the rolling ocean billows dashed unceasingly against the steep face of the reef and sent great seething sheets of roaring foam sweeping shoreward over the surface of the coral table. Yet all along the edge of the reef were numbers of women fishing with rods. Sometimes, when a roller too big to withstand rose and curled its greeny crest fiercely before them, the women would run landward a hundred feet or so, and let it sweep by them waist high. Then they would hurry back to

the face of the reef and drop their lines into
the sea again. At one place, where the curve
of the reef broke the first force of the rushing
seas, were gathered some dozen or so of young
girls, all standing up to their waists in the
troubled surf, catching a species of small rock
cod that came in with the rising tide, and
dropping them into the baskets carried on their
naked backs. Every now and then, however,
a wavering, leaping wall of backwash from the
shore would make them spring for safety upon
the round, isolated knobs of coral that here and
there studded the ledge on which they stood.
For any one of them to lose her footing meant
being carried out by the backwash over the edge
of the reef, and, if not drowned, being severely
lacerated by the jagged coral. My friend and
I sat looking at them for some time, when presently he said :

"Look at that girl right on the very edge—
the one with the big bundle on her back. She'll
get knocked over by the next sea to a dead
certainty. By jove, it's a child she's carrying.
Man, it's your youngster ! "

In another moment we tore down the rocky
path, and plunging into the water, ran along

the reef, falling into holes every now and then, and clambering out again, half-drowned. The group of girls saw us coming, and gave the wicked E'eu a warning cry. She turned, and at once made off for the shore, making a *détour* to avoid the vengeance she knew was coming. She gained the beach first, barefooted as she was, and, dropping her rod, sprang up the path like a goat. My child, I saw, was strapped tightly upon her carrier's back by two cross-belts of green *fau* bark, which came up under her (E'eu's) armpits and met across her naked bosom. That the infant was drenched through and through I could see, for her white linen sun-bonnet hung flat and limp over her little head.

We stopped to break off a couple of stout switches and started in pursuit, reaching my door just as E'eu darted into the inner room to disencumber herself of her burden.

Two minutes later we haled her out on to the sward, and Hakala gave her a beating that she will remember to the end of her life. This sounds brutal, but it was the only way of making her understand the nature of the shock my parental feelings had sustained ; a cuff or two would have done more harm than good. Half

an hour later the town councillors deputationised
me again, and suggested that I should invest
E'eu with the order of the sack and replace her
with Pépé. I said I would give her another
week's trial. After the deputation had left I
told Hakala to bring E'eu into the house, so
that while she was still feeling the effects of her
whipping I might impress upon her youthful
mind the enormity of her conduct. But she
could not be found. Presently, however, we
heard the murmur of voices from the cookhouse.
Walking softly over, I peeped in through the
window. The place was in semi-darkness, but
there was still enough light to fill me with
wrath at what I saw. There, stretched upon
the floor, face down, was the under-nurse,
supporting her chin upon her hands, a cigarette
in her mouth, and that villain of a Moemoe
lubricating her glossy brown back with a freshly-
opened tin of my Danish butter, into which he
now and again thrust his fingers. The extreme
annoyance I manifested seemed to astonish Moemoe, who protested that he had used but very
little butter, and Hakala had refused to lend
him her bottle of scented hair-oil.

But all these were minor troubles. Every

day people—principally women and girls—from
Fatiau, and Hakupu, and Alofi came outside
my compound wall and challenged my female
domestics severally and collectively to come out
and fight ; and thereby made my life a misery.
Pépé the Rejected was, I think, the instigator,
and at last I had to compromise matters with
her by presenting her with a cheap concertina
and a tin of salmon. After that the personal
assaults ceased, and the opposing forces merely
indulged in vituperative language across the
compound wall.

Gente Hermosa: the Island of "Beautiful People."

L ONG, long ago, when Fernando Quiros, the Spanish pilot, sailed among the islands of the great unknown South Sea, his ship touched at a small, low-lying island, densely covered with groves of palm trees which grew to the very verge of the bright, shining beaches of snow-white sand that encircled them like a belt of ivory. This island was inhabited by a race of light copper-coloured natives of "gentle demeanor and very great beauty of person;" in fact so ravishingly beautiful were many of the females who boarded the ship that Quiros, in the chart he sent home to his royal master in Spain, named the island "Gente Hermosa"—handsome or beautiful people. There is much doubt, however, that the Gente Hermosa of Quiros is the Gente Hermosa (now

called Quiros Island or Swain's Island of our
modern charts) of his voyage with Torres,
situated about a day's sail northward from
Samoa. That is but a tiny coral gem about
three miles broad and one in width, and
containing less than 3,000 acres, for the centre
is taken up by a shallow, salt water lagoon.
But though so small this lovely little isle has an
interesting history, which, however, need not be
told here, its name being only mentioned lest it
might be imagined that it is the island of
" beautiful people " spoken of by Torquemada,
the historian of the voyage of Quiros.   The
true Gente Hermosa is some distance from there
—in 10 deg. 2 min. south, and 161 deg. 10
min. west (while the Gente Hermosa or Swain's
Island of the modern charts is in 11 deg. 10
min. south and 170 deg. 56 min. west) is called
Rakahaaga, and lies about twenty miles N.N.W.
of the beautiful island of Manhiki, one of the
loveliest spots in the Pacific.   The late Major
Sterndale, than whom there was no better
authority upon the manners, language, and mode
of life generally of the Malayo-Polynesian race,
spent a considerable portion of his time during
his island wanderings on Rakahaaga, and know-

ing both it and the Gente Hermosa of the charts
equally well, gave it as his opinion that Quiros
had assigned Rakahaaga a wrong position on
the chart he sent home to Spain. In the first
place Rakahaaga has always, or at least
within the last 200 years, carried a population
of over 500 people, while Swain's Island was
uninhabited up to within fifty years ago, and
although there were found a few stone hatchets
which showed human occupancy at some long
past time, there were no signs of the great
depressions and high banks resulting from the
construction of artificial swamps for the cultiva-
tion of taro. The presence of these certainly
would have proved that the island had sustained
a permanent population, even though they had
died out or been killed a hundred years back.
It is, therefore, very evident from this fact
alone that the old cartographers of Quiros's
voyage were in error, and that the island of
" beautiful people" was the now well-known
Rakahaaga.

At the present time the population is something
over 500. Fifty years ago, when it was visited
yearly by many ships of the American whaling
fleet, there were, it is said, over 3,000 natives

in Rakahaaga and Manhiki. The steady and
lamentable decrease in their numbers is ascribed
by the missionaries as being the result of their
intercourse with the seamen of the whaling fleet,
and from the fact that the women of both
Manhiki and Rakahaaga were much sought
after by wandering white traders from all parts
of the North and South Pacific ; but Captain
English, of Fanning's Island, and Major Sterndale, both able authorities, ascribe their
decadence to the introduction of European
clothing by the missionaries, which, as in
numberless other instances in the annals of
missionary enterprise in Polynesia, was attended
by the most disastrous results.   By some incom-
prehensible fatuity, the earlier missionaries in
the Pacific were imbued with the idea that these
people, who for centuries had worn nothing
beyond a girdle or waist-cloth of native
manufacture, could at once adapt their constitu-
tions to such a violent and radical change as
that caused by their clothing themselves in
heavy woollen garments sent out from England
by those interested in the spread of the Gospel.
Not even the fearful consequences that attended
the clothing of the people of Raratonga early in

the present century by the enthusiast Williams could teach them sense in this respect ; and the natives themselves, though now accustomed to the use of European costume from their earliest childhood, assert that the ravages of pulmonary disease, to which they are particularly liable, first began to decimate them after they accepted Christianity.

Without doubt the Rakahaaga natives may claim to be one of the handsomest races in Polynesia. Their complexion is very light, and their smooth, glossy skins are not now disfigured by tattooing, except in the cases of the men, who have been tattooed at other islands when voyaging about the Pacific in whaleships. The younger women and girls are perfect models of symmetry of form, and their large dark and languishing eyes, oval faces and pearly teeth, and rosy flush of cheek under their clear skin, at once give reason for their being so sought after as wives by the old-time traders. In their dispositions they are bright and cheerful, and both men and women seem passionately devoted to children, and mingle with them in their childish games in a manner that at once impresses the beholder with a very high estimate of their gentleness and amiability. The men are all expert divers, and many of them display great skill in various handicrafts, such as carpentering, boat and house building, etc. For their knowledge of such they are indebted to Captain John English, a veteran and honoured old trader, who, looking for suitable labourers to work at his coconut oil manufactory of Fanning's Island in 1860, visited Rakahaaga and Manhiki and took some hundreds of them away with him to that isolated atoll. The aptitude they displayed in learning the arts of civilisation was truly wonderful, and the old captain took a natural and sincere pride in his pupils, and on their return to their home some three years later they effected a radical change in the condition of their fellow-countrymen by teaching them all the knowledge they had acquired during their sojourn with the old sailor.

The houses are built of coralline, and are constructed in a neat and substantial manner. Each dwelling is enclosed within a low wall of plastered coral stone and sand, and most have European doors and windows, constructed from native or imported timbers by the people themselves. The floors are covered with beautifully

worked mats of handsome design and fine
texture, and in some of the dwellings of the
better-off people may be seen good European-
made furniture, such as chests of drawers, etc.
Of late years the adoption of European clothing
has become general with both sexes ; but the
black suits, chimney-pot hats, and Sarah Gampian
umbrellas of the early missionary days have
very sensibly been discarded for clothing of
a lighter texture, supplied by the resident trader
at Manhiki, and the occasional trading vessels
that visit the two islands. They are all
Christians, and, so far as outward observances
go, could be, and indeed are, held up by the
missionaries as bright and shining examples of
the beneficent results attending the introduction
of Christianity. But, like too many such island
communities, their Christianity is but a garment
to be put off and on at their own convenience ;
yet despite this they are on the whole an
amiable, interesting, and well-mannered people.
Contrasting their condition of peace and plenty
with that of so many thousands of English
labourers and artisans of the present day, one
cannot but envy their happy and contented
existence.

Deep-Sea Fishing in Polynesia.

WITH the exception of the coast of New Zealand I do not think that there can be better deep-sea fishing grounds in the whole Pacific than the calm waters encompassing the many belts and clusters of the low-lying coral islands of Polynesia. Unlike the fortunate inhabitants of such mountainous but highly fertile groups as Samoa, the Society, Cook's, and Austral Islands, the people of these low, sandy atolls literally depend upon the sea for their existence ; for, beyond coconuts, the drupes of the pandanus palm, and a coarse vegetable called puraka (a species of gigantic taro), they have little else but fish to support existence. The result of these conditions is that they are very expert fishermen and divers, and the writer, during a twenty-six years' experience of the Pacific Islands, was often lost in admiring wonder at their skill, courage, and resourcefulness in the exercise of their daily task of fishing, either in shallow water within the reef, or miles away from the land on the darkest nights, and using tackle of such weight, size, and peculiar construction that the uninitiated beholder imagines he is living in the age of primeval man.

Flying-fish catching is, perhaps, the most exciting and exhilarating of all South Sea Islands' fishing. There are many methods ; for they are caught both by day and night. In the Gilbert and Kingsmill Islands, lately annexed by Great Britain, their capture by day is effected by thin lines trailed astern by sailing canoes, and at night by the aid of torches with nets. In this latter manner, however, the light-skinned inhabitants of the Ellice Islands surpass their northern neighbours of the Gilberts. Let us imagine that just about sunset we are on one of the low-lying islands of the former group.. The mission bell, or rather the wooden cylinder that does duty for such, has been rung for evening prayers. The evening is calm and quiet, and the smooth sea gives

9

promise of a good night's sport, so the prayers
are hurried through as quickly as possible.
Ioane, the Samoan teacher, is an ardent fisherman himself, and the moment prayers are over
he whips off his white shirt and pants, and ties
a *titi*, or girdle of leaves, around his loins.
Outside, and all along the white beach, numbers
of men, women, and girls are hurrying to and
fro, getting the canoes ready.   Upon the frame-
work of each outrigger is placed a bundle of
huge torches made of the dried leaves of the
coconut palm.   These torches are from 6ft. to
10ft. in length, and are very easily and rapidly
made by the women.   So that they may not
be wetted by a sea breaking over them as the
canoe crosses the reef, they are covered over
with a coarse mat of coconut leaves, called a
kapau.   Each canoe takes four, or at the
outside five men, two to paddle, one to steer,
one to act as torch-bearer, and another to wield
the scoop-net.   All being in readiness, the
canoes are lifted up, carried into the water, and
the fishermen take their places in them, paddles
in hand.   Sometimes there are—even on little
Nanomaga, one of the Ellice Islands, and which
has only a population of 400 souls—as many as

thirty or forty canoes engaged. Watching for a favourable opportunity, when a lull occurs in the break of the surf on the reef, the paddles are struck into the water, and the little fleet dash over into the deep water beyond. Upon each outrigger gleams a tiny spark of red fire. This is from a small torch made of the split spathe of the coconut, and is used to ignite the larger torches at the fitting moment.

The utmost silence prevails. Pipes are lit, and then a whispered consultation takes place. The fleet divides itself, one portion paddling slowly along the reef towards the south end of the island, and the other to the north. In a quarter of an hour each party has reached the position assigned to it. They are to work towards each other. Every now and then a flying-fish leaps out of the water and flies away into the encompassing darkness, to fall with a splash hundreds of feet off. At last the canoes are in line, forming a semi-circle.

Ta ! ("Light up") is the steersman's whispered order.

The man who has hitherto paddled amidships lays his paddle down quietly and stands up, one foot on each gunwale, and a torch is passed up

to him. Thrusting the burning end of the
smaller torch into the crisp dry leaves, he blows
steadily for a moment or two till they burst out
into flame, and at the same instant from twenty
other canoes a glare of light illuminates the ocean.

The moment the big torch is lit a native
takes up his position in the bows.   He holds in
his hands a light pole of *pua* wood about 14ft.
in length, to one end of which is attached a
round scoop-net stretched upon thin bent wood.

By this time all the torches are burning
bravely, and the canoes advance.   The bright
flame reveals the surface of the water so clearly
that the smallest speck upon its surface can be
discerned.   A quarter of a mile away on the
starboard hand the white line of breakers rear
and tumble upon the jagged ledges of black
reef ; beyond lies the dim line of beach fringed
by the long line of sleeping palms.

Presently the torchbearer of one of the
canoes on the seaward horn of the crescent
gives a warning " hist ! " and, giving his torch a
shake, a shower of sparks fall.   The flames
burst out brighter than ever, and, as every
other torchholder follows his example, the
paddlers and steersmen look over the side and

the men forward hold their nets in readiness, and their keen eyes search the water ahead and on each side of them. In another moment the canoes are in among a swarm of flying-fish, lying almost motionless upon the surface, dazzled by the blinding light.

With quick, lightning-like sweep, the man forward darts his scoop upon the water, and, swiftly reversing it, raises it again, and a cry of triumph escapes him. *Maté* (struck!) he cries. Inside the netting there is a gleam of burnished silver and blue, and, with a deft sweep, round comes the scoop, and half-a-dozen fish are dropped into the after part of the canoe, where they lie beating out their lives with outstretched fins and quivering tails.

There is no need for silence now, and shouts and cries of emulation from one canoe to another reach the ears of the women on shore, who call out in return. Splash, splash, and the deadly scoops sweep up the fish which lie grouped together in dozens, and then, realising at last the fate that awaits them in that fierce, strange light, they leap and fly hither and thither in every direction, and seek refuge in the darkness.

Then, as if by magic, the sea is dark again,
and in place of the loud laughter and shouts, a
sudden silence.   Every torch is out, the paddles
are taken up again, and for a few minutes the
canoes paddle quietly along with noiseless
stroke.   Away towards the northern end of
the land a bright circle of lights appear, and
faint cries are borne across the water as the
other half of the fleet plies its scoops among
the fish.   Another ten minutes pass, and then
again the torches of the southern fleet flame
out, and the scoops strike the water and swirl
about ahead, astern, and on each side.   At last
the leading man calls out :

"It is enough, friends.   Our canoe is half
filled.   How is it with thee ? "

"Enough, enough !   Let us now fish for
tau-tau."

So off they paddle shoreward, where the
waiting women meet them on the beach, and
fill their baskets with flying-fish.   Then, keeping perhaps twenty or so for bait, each canoe
once more paddles out into the darkness to fish
for *tau-tau, palu,* and *takuo* till the morning.

Let me attempt to describe what these fish
are like, and the method in use for catching them.

The *tau-tau* is exactly like an English salmon as far as appearance goes, save that the entire head is formed of a series of hard, long plates, and the jaws are fitted with teeth that resemble those of a rip-saw. The *palu* is caught only in deep water and varies from 2ft. to 6ft. in length. It has a dark brown skin, is scaleless, but covered with minute horny projections that curve outward and backward like the feathers of a French fowl. Its flesh is greatly relished by the natives, both as a food and for its highly medicinal qualities in some complaints. It is, however, of such an oleaginous character that it is only eaten alone when needed as a purgative ; generally it is mixed with beaten-up *puraka*, and is very palatable, even to European tastes.

The *takuo* is a species of gigantic albicore, and I have caught them in both the Caroline and Marshall Groups in the North-West Pacific up to 120lb. weight. Like the *palu* and *tau-tau*, it is of semi-nocturnal habits, and is seldom caught but just before the dawn, or an hour or so before sunset.

Preying as they do unceasingly upon the swarms of flying fish that hover about the coral reefs of the Equatorial Islands, the *tau-tau* are

themselves preyed upon by a fish of immense
size, called the *pala*—a fish quite distinct from
the *palu*, for it is surface-loving and is never
caught at night.   These sometimes attain a
length of 7ft., but in girth they are comparatively slender.   In consequence of their
flesh being somewhat coarse, they are not as
much esteemed by the natives as the *tau-tau*,
and in times of fish-plenty are seldom captured.
They are a beautifully coloured fish, with dark
blue backs, and sides of silvery grey, and in
shape exactly resemble the English mackerel.
The head is like that of the *tau-tau*—a series
of hard bone-plates—and the jaws, which in a
full-grown fish are a foot in length, are armed
with a single row of terribly sharp, saw-like
teeth, capable of severing a man's arm as easily
as if it were struck off by a blow from an axe.

The natives of the little island of Nanomaga,
in the Ellice Group, six hundred miles from
Samoa, are famous for the skill they display in
pala fishing, which is of a very exciting nature,
on account of the great strength and swiftness
of even a moderately-sized fish.   The tackle
employed is made of four-plait coconut cinnet,
and the hook used is always of European make

—generally a small shark hook, purchased from the local white trader. If the shank of the hook is not of sufficient length, the line for about 3ft. up is seized round with wire, so as to prevent the *pala* from biting the line and escaping. A whole flying fish is used for bait, and the time chosen for the sport is the dawn. Five men generally form the complement of a *pala* canoe, four to paddle, and the fifth (who acts as steersman as well) attends to the line. At a distance of a mile or so from the land, pieces of flying fish are thrown overboard at intervals, as the canoe is urged swiftly along. A swirl and splash astern denotes that a *pala* is following the canoe and devouring the pieces of fish. The baited line is then dropped astern and allowed to trail out for, perhaps, 50 yards or so, the paddles urging the canoe to increased speed, for the greater the speed the better chance of the fish seizing the bait. Another splash much nearer the canoe, and the steersman, dropping his paddle, takes the line in both hands, and hauls it in as quickly as possible, while the four paddlers urge the canoe along at topmost speed. Suddenly the steersman feels the stout line tauten, and at

the same moment a tremendous splash astern
shows that the prize is hooked.

In an instant the paddlers reverse their
positions, and face toward the man holding
the line.   The bow of the canoe has now
become the stern, and the man that has
previously paddled on the forward thwart becomes the steersman.   For the *pala* is strong,
and gives a canoe a long run before he is
exhausted enough to let a bowline be slipped
over his lengthy body.   Away goes the canoe
with the speed of a bird on the wing, and
the man holding the line takes a turn round
the thwart to prevent the rough cinnet from
cutting his fingers.   Suddenly the strain ceases,
and the line is pulled in ; the *pala* has paused.
But only for a few seconds, for his great
crescent-shaped tail rises out of the water, and
then he " doubles," or " mills " as whalemen
say, and darts back at a terrific speed.   With
deft hands the line is passed to the other end
of the craft again—for there is no time to
slew a canoe when fastened to a *pala*—and
away she goes.   Fortunately, the sea is generally smooth ; were it otherwise the canoe would
ship so much water that it would be swamped, or

the line probably part from the additional strain.
After another run of half a mile or so the speed
begins to decrease, and the line is steadily hauled
in till the gleaming blue and silvery body of the
prize is plainly visible as he rolls from side to
side, and, salmon-like, shakes his jaws in a fierce
effort to free himself from the hook. Foot by
foot the big brown-skinned native who holds
the line hauls it in, and his keen eyes watch
every movement of the fish; for sometimes a
pala when almost exhausted will yet have
strength enough remaining to dive under the
canoe, and foul the outrigger, which would
mean a capsize and his eventual escape from
his pursuers. At last he rises to the surface,
swimming still, but lying first on one side,
then on the other. Then the man amidships
hands the line holder a short but heavy and
rudely fashioned club much used in despatching
sharks.

Passing the line along to the next man, he
goes forward to the head of the canoe, and
crouches down, club in hand, while the line is
steadily hauled in, care being taken by the
steersman to get the fish on the port side,
so as to avoid the outrigger, which it

might foul or break in its dying struggles. At last, drawing back his brawny right arm, the man deals the fish a heavy blow upon its bony head. There is a terrific splashing and leaping of foam, the *pala*, half-stunned but still strong, describes the segment of a circle in his dying flurry, and his broad, crescent-like tail strikes the water with resounding smacks. For a minute or so he lies quiet, and then, with a last, short struggle, gives up the fight, and is hauled alongside.

Often a *pala* is of such great length and weight that he cannot be either put into the body of an ordinary small canoe or placed upon the platform of the outrigger without great danger of upsetting the craft. In such a case he is either cut in halves and lifted into the canoe, or towed alongside the canoe. The latter course is not unattended with danger from sharks, who, rushing at the fish, either carry it away bodily, or stave in the canoe in their efforts to do so.

The Tia Kau Reef, a few miles from Nanomaga, which is alluded to elsewhere in this work, is without doubt one of the best fishing grounds in the South Seas. It is a submerged coral platform, and in the shallowest part has a depth of about four fathoms. It is about two hundred and fifty acres in extent, and from its apex the sides gradually slope in a graceful rounded curve, till its outlines are lost in the blue depths of the ocean. During the ordinary trade-wind weather a ship may sail over even the shallowest part in perfect safety, but during the prevalence of the westerly winds —from January to March—the sea breaks over and around it to such a degree that it resembles a vast cauldron of white, roaring foam. Upon and around this spot fish of every size, shape, and colour swarm in prodigious numbers. But hordes of voracious sharks infest the place, and render fishing terribly dangerous after nightfall, and, indeed, occasionally in daylight. The people of Nano-maga occasionally form a large fishing party to the Tia Kau, and the writer, who has often accompanied them, has seen as many as twenty canoes loaded to sinking point with an extra-ordinary variety of fish in less than an hour. Among those caught was a species of red rock-cod of huge size and appalling mouth. Three of these would fill a canoe, for, although none of them exceed 3ft. in length, they are of a tremendous girth. Leather jackets of every imaginable hue, shape, and size rush the hook, and actually come to the surface and knock against the canoe. Some of these, with light blue skins and red eyes, grow to a great size— 1olb. to 2olb.—and carry a serrated spike on their backs a foot in length, and as thick as a man's finger at the base. Grotesquely ugly as is their shape, they are, however, of a very delicate flavour, and the natives are very fond of this particular species of the *isumu moana* (sea rat) as they very appropriately term leather jackets generally. Darting to and fro in every direction upon the surface of the water, great garfish, of the kind known in Australia as "long toms," will try to seize the hook before it can descend to the coral bottom. Curiously enough these fish, while of a very delicate flavour, are not at all relished by the natives, who call them "foul-feeders"—*i.e.*, feeders upon dead bodies. Upon the top of the reef vast swarms of a fish called *tafau uri* (a species of trevally) almost hide the gorgeous beauties of the coralline growths below, and mingling with them appear an extraordinary variety of parrot-fish and other rock-haunters. Here and there, in the deep holes and strange, weird-looking chasms that everywhere break open the crust of reef, deep down below, hawk-bill turtles swim lazily to and fro, and above them prowl the great grey ocean sharks, waiting for the night to come, and turning lazily about, with green, malevolent eyes, to watch some turtle, more daring than his fellows, venture out from the protection of his rocky hole or cavern. Then, besides the big grey sharks, there are others—blue, hammer-heads, and the savage little white and brown tiger shark. These give more trouble to the canoes than the big fellows, for they are so swift and active that once they assemble beneath a canoe it is impossible to haul a fish up through them.

Only in the daytime is it possible for any one to fish on the Tia Kau ; at night it would simply mean risking a horrible death from the sharks for any one who would be rash enough to attempt it alone in a frail canoe ; and even in a stout boat fishing in such waters after dark is by no means unattended with risk.

Birgus The Robber: The Palm Crab of The Pacific Islands.

THE palm crab, or robber crab, as it is also called (*Birgus latro*), is certainly an ugly fellow, but must not be judged entirely by his outward appearance. In the South Seas we especially admire him when, in place of the blue and grey tints that were his in life, he has changed in the oven or pot to a lovely lobster-like red. In addition to the names above mentioned, he is known also to travellers as the " coconut-eating crab " and the "giant land crab," while to the white traders and natives of the South Seas he is simply *ūū* or *Uga po*, "the night crab," something that one can smack one's lips at when your native cook disinters him from his mummy-like bandages of coconut leaves and reveals him with body of bright red shell and pendulous tail of blue fat. For, hideous as he appears, the palm crab of the Paumotu, Society, Tonga, and Ellice Groups in the South Pacific is a creature greatly prized by both natives and whites for the delicate flavour of its white flesh, and for the fat of its short lobster-like tail. Among the high, mountainous, and forest-clad islands of both the North and South Pacific, it is not so common as among the narrow and sandy palm-clad atolls of the Ellice, Tokelau, and other equatorial archipelagos, where, on account of the paucity of human inhabitants, and the enormous number of coconut trees and pandanus palms which afford them food, these crabs are very plentiful, and attain a great size. Especially is this so on the sparsely inhabited but noble lagoon islands of Christmas, Mādurŏ, and Fanning's Atolls, where, with coconuts, turtle, fish, and sea birds, they constitute the food of the brown-skinned, stalwart natives.

The prevailing colours of the *ūū* (pronounced *oo-oo* by the Samoans) are various shades of grey and blue on the carapace of its back and tail, and a pale yellow or white underneath the whole of its body and legs, changing into a light red as the last joint of each leg, or rather claw, is reached. Nocturnal in its habits, it is but seldom seen during the daytime, and then the sound of a footstep sends it hurrying through the underscrub, or over the loose coral slabs that line the margin of the lagoon, to its burrow, which may either be beneath the roots of a large tree, or far down below a pile of coral stones banked up by the action of the sea. During our ship's stay at Arrecifos Lagoon, in the North Pacific, our boat's crew of natives, however, captured a good many in the daytime by placing a number of pieces of broken coconut in the pathway near the crabs' retreat overnight, and watching for their appearance at daylight, when they were returning from their nightly wanderings on the beach. Their capture was effected by slipping a noose of cinnet over one of the huge nippers, and then winding it dexterously over and over his claws and body till the creature was securely bound. Very strong cinnet must be used, for so great is the *ūū's* strength that it can snap an ordinary fishing line like a piece of cotton thread. Although omnivorous in its diet, the giant land crab subsists principally upon the rich saccharine drupes of the pandanus or screw pine and the delicate pulp of young coconuts. In almost every palm or pandanus grove, particularly those which are situated upon the smaller islets of an atoll, the evidences of their occupation may be seen in the vast quantities of young coconuts lying upon the ground with a large hole torn through the husk, and an empty interior. Gregarious to a certain extent, the robber crabs yet each have their separate burrow, and resent the intrusion of another of their kind most fiercely. Whilst living at Arrecifos we had ample opportunity of studying their habits, and found that, although so shy and timid in the light of day, they sallied forth at night time to feed with the utmost confidence, even though many people were about. Like the flying fish, they seemed to be unable to resist the attraction of a light at night, and our native sailors, providing themselves with torches of dried coconut leaves, would sometimes capture two or three hundred at once by means of dragging a turtle net along the beach. During those months of the year when the green turtle come ashore to deposit their eggs in the sand, the robber crab has a good time, and devours quantities of this delicate food,

Its curious protuberant eyes, that enable it to
see all round the compass at once, soon discover
a place where the sand has been disturbed by
a turtle, and, travelling along with an absurd,
stilt-like motion, it reaches the spot, settles
down to work, and, with nippers thrown out
upwards and backward, begins to probe the sand
with its sharp, powerful, and serrated claws till
it feels the eggs, the tough skins of which it
rapidly tears open and then feeds upon the yolk.
One morning a native sailor came to me with
an amused face, and asked me to come with
him along the beach a little way, as he wanted
me to see something very curious.   This was
a turtle's nest, as yet undisturbed, but on it lay
three huge robber crabs locked together in a
deadly grip, and quite incapable of freeing
themselves, though their curious glassy eyes
moved to and fro in alarm at our presence.
One of them had three or four of his
armoured legs crushed to a pulp by the powerful
nippers of one of the trio.   Evidently they had
all reached the nest at the same time and there-
upon engaged in combat.   My companion soon
secured the lot by tearing up a long vine that
grew near and binding it round them.

The oft-repeated statement that the robber crab ascends the coconut tree and there husks the nut is only a traveller's story, although, if it cannot find a fallen nut, and other food be unobtainable, it will, if it can discover a short coconut tree bearing fruit, ascend it, and, nipping the stalk, let it fall. It will then descend to the ground, and proceed, not to tear off the husk, but to make a hole in the soft, fleshy part of the top, and thus reach the interior. At the same time a full-grown robber crab, if it cannot find a young nut, makes no difficulty of tearing off the husk of an old, fully-matured one. Bit by bit it strips off the tough, wiry covering until the " monkey's face " is revealed. Into the " eye " of this it inserts the tip of one of its sharp claws, and works out a space sufficiently large to at last permit it to begin operations on the hard shell with his nippers. Then it snaps away piece after piece till the orifice is large enough to allow it to clean out the entire nut, which is left scraped of every remnant of pulp.

The fatty, blue-coloured tail of these creatures is esteemed a great delicacy by the natives of the equatorial islands of the Pacific, and the oil from it certainly is, as the writer knows by experience, a wonderful remedy for rheumatism. When the fruit of the pandanus tree is fully ripe, and the rich, juicy drupes fall upon the sandy soil, the robber crab thrives, and its heavy tail swells out with the fatness thereof till it appears to be as much as it can do to carry it.

The immense muscular power in the two great nippers of a full-sized robber crab renders the greatest caution necessary when one is being captured. If by any mischance one's hand should come within reach of his grip, the bones of whatever finger or fingers are seized would be crushed. I well remember one such instance in Funafuti Lagoon, an island of the Ellice Group in the South Pacific. A young native employed by an old white trader there one day accompanied his master to a little islet on the west side of the atoll, to search for *ūū*. A quantity of teased-out coconut fibre and a pile of coconut shells at the foot of a large pua tree indicating the burrow of one of unusual size, the native searched for and soon found the hole leading to the lair. Cautiously inserting his arm up to the shoulder to feel in which way the tunnel led so that he might dig the creature out, he was suddenly seized by the wrist. Suffering the most acute pain he tried to drag his arm out, but the size and strength of the crab proved too much. His loud outcries were fortunately heard by the white man and also by a party of women and children who were fishing on the reef close by. Rushing to his assistance they at once tore away the soil and net-work of roots till the *ūū* was reached and a sheath knife was plunged through its pendulous tail. The crab at once released its hold of the wretched man, who, however, was severely injured, and never recovered the full use of his right hand again.

In captivity the robber crab, when fed on a diet of coconut, with an occasional bit of raw fish, thrives very well, and becomes perfectly resigned. An empty ship biscuit tin is selected for its prison and proves a secure one unless there should be a hole or crack anywhere along the sides or in the bottom. This the " robber " would soon discover, and inserting the tip of a claw into the place, tear the tin apart like paper and make its escape.

To the above account which originally
appeared in *The Field*, the editor of that
journal added the following :—It is not without interest to supplement the above account
with a note on two points that have been
overlooked, and well deserve mention. The
first is the singular modification of structure
which this species of an aquatic family has
undergone to fit it for a prolonged existence on
land. Professor Semper has pointed out in his
admirable work, " The Natural Conditions of
Existence as they affect Animal Life " (Inter-
national Science Series, Vol. xxxi.), that following
on its change of habit a portion of the gill cavities
have become modified into an organ for breath-
ing air, namely a lung. This is described and
figured on pp. 5 and 185 in the work referred
to. The other point is the use to which the fat
is put by sagacious travellers. Mr. H. O.
Forbes, in his " Naturalist's Wanderings in the
Eastern Archipelago," writes : " It accumulates
beneath its tail a store of fat which dissolves by
heat into a rich yellow oil, of which a large
specimen will often yield as much as two pints.
Thickened in the sun, it forms an excellent
substitute for butter in all its uses. I discovered

it to be a valuable preserving lubricant for guns, and steel instruments ; and only when a small bottle of it which I had for two years was finished, did I fully realise what a precious anti-corrosive in these humid regions I had lost."

On an Austral Beach.

A S we sat, half asleep, on the shady verandah of the local public house (otherwise the " Royal Hotel ") listening to the ceaseless pounding of the surf on the ever-restless bar, a dusty, slouch-hatted horseman rode along the baking street, pulled up when he saw us, and, in a voice indicative of a mighty thirst, besought us to have a drink with him. We consented, and then Sandy Macpherson—that was his name— gladdened our hearts by telling us that he and his mate wanted us to come out to their camp on the ocean beach for a couple of days' fishing and shooting. The two men were beach- mining, that is, working the deposits of auri- ferous sand that are to be found all along the north coast of New South Wales, and their camp was situated on the margin of a tidal lagoon at less than fifteen miles away from the dull little township where my friend and myself were awaiting the arrival of a steamer to take us to Sydney.

By daylight next morning we had breakfasted and saddled up, and long before the inhabitants of the sleepy old seaside town were awake, we cantered through the silent streets and out along the winding forest road leading towards the coast. Five miles from the town we emerged from the gloomy shadows of the grey-boled gums out upon the summit of a hill whose seaward side, clothed with a soft green nap of low shrub as smooth as an English privet hedge, was shining bright in the first rays of the morning sun ; while at its base of black trap-rock the lazy ocean swell had not yet heart to break, for only the lightest air ruffled the surface of the sea beyond. To the north, cape and cape and headland and headland of pale misty blue were fast purpling in the glorious sun, and southward there trended in a great sweeping curve a noble beach full fifteen miles in one unbroken stretch. Beneath us, where it began, its hard surface at the water-margin was dotted with countless groups of snow-white seagulls

and jetty plumaged "redbills," all standing or
sitting in motionless array with heads facing to
the sea ; further on the line of beach was yet
but half revealed, for the smoky haze of a semi-
tropic sea still hovered o'er it, and floated in
gossamer-like clouds up towards the dark green
fringe of scrub-clad hills.   Oh, the beauties of
a summer's morn upon that wild and lonely
coast !   The strange, sweet, earthy smell of the
rich red soil beneath our horses' feet, the sweeter
calls and cries of awakening bird life around us,
the glint of the blue Pacific before and the dome
of cloudless turquoise above, and the soft cool-
ness of the land breeze as it stole gently seaward from the mountains, and stirred and
swayed the leafy banners of the lofty gums and
tapering bangalows !   And then, as we turned
and followed the bearded Mac adown the
narrow fern-lined track that led us to the
shore, the blue sky above us vanished and
showed but here and there through the thickly
overarching branches and clustering vines and
serpent-like lianas ; a big black wallaby leapt
across the path just in front of our leader's
horse, then another and another, and all three
crashed into the thick undergrowth of the sea-

ward hill slope ; a flock of green and golden
parrots shrieked angrily at us from the boughs
of a blossoming honeysuckle—and then we came
out again into the light and warmth of the sunshine and the noise of the tumbling surf and
seabird clamour of the open beach.

" Now we can go as hard as we like," said
Sandy, and away he shot before us over the
hard, firm sand, riding close to the water's edge,
and hurrooing wildly at the whirling clouds of
seagulls and divers as they rose with hoarse,
protesting croaks at the galloping feet of our
shoeless steeds. Two miles onward, and then a
tiny shining stream of water that cut its way
through the sand to the sea brought us up
sharply—for our horses knew the danger of a
quicksand—and we walked cautiously up to
beyond high-water mark, and there crossed.
Then, for the next hour or so we walked or
trotted soberly along, smoking our pipes and
watching through the transparent green of the
rollers as they curled to break fifty yards away,
the darting forms of countless thousands of
great sea-mullet swarming beachward with the
rising tide. Sometimes as we approached too
near to the water there would be an agitated

swish and swirl and bubble, and a compact body
of keen-eyed " blue-fish " or whiting that were
cruising in water scarce deep enough to swim in,
fled seaward with lightning-like rapidity.   That
we should see plenty of fish along this beach we
knew, but we were not prepared for the extraordinary sight that we witnessed a mile or two
further on ; for here, at the mouths of two
little creeks which ran down to the sea in
parallel lines not a hundred yards apart, the
water was literally teeming with countless thousands of silvery bream, trevally, whiting and
garfish, and every wave seemed to add to their
number.   Swimming so close to the shore that
every now and then some hundreds would be
left stranded on the sand by the backwash, were
swarms of dark green-backed garfish, about
fifteen inches long.   It did not take us long to
discover the cause of this gathering of the clans
—the banks of each rivulet were covered with
layers and ridges of fine big prawns, and by the
turmoil at the creeks' mouths it was evident
that numbers were being swept down into the
hungry jaws that awaited them.   On the previous evening, so Sandy told us, there had been
a violent thunderstorm, the creeks had risen

suddenly, and rushing down to the sea had met
the incoming spring tide, and every wave that
broke upon the shore left some hundredweights
of stranded prawns behind it, where they remained to be devoured by the gulls and divers
and the vast bodies of fish which, when the tide
again rose, were enabled to ascend to the higher
parts of the beach. The greater number of
these delicious crustaceans were still alive, only
those which had been washed apart from the
thicker masses and ridges of the others having
succumbed to the rays of the sun. Sandy had
come prepared, for he now dismounted and
began to fill a sugar-bag with them. Prawn
soup, he told us, was " verra good." Also he
observed that if he could only get but one-half
of these masses of prawns up to Sydney market
they would be worth £25 to him, and that was
more than he could make by three months'
hard work at beach-mining. Before we mounted
again we watched our chance and picked up
nearly a dozen of stranded garfish ; these
Sandy popped into the bag on top of the
prawns.

For another mile or two we rode slowly along
the hard, unyielding sand, till we came in sight

of a high sand head—the northern side of the
entrance to a tidal lagoon—and then turned
away up from the beach to where a thin, wavering line of smoke ascended from the scrub.
This was the diggers' camp, and in a few
minutes Sandy's mate came to meet us, carrying
a couple of fat black ducks which he had just
shot. The tent was situated on a little grassy
bluff that overlooked the lagoon, and while Sandy
and his mate set about to cook our breakfast of
grilled garfish and grilled duck, my companion
and I took their guns and set out along the
lagoon bank, taking care to walk with discretion, for black snakes were very plentiful. A
ten minutes' walk brought us to an open space,
from where we had a splendid view of a lovely
scene. The broad, shallow lagoon, with its
shores lined with she-oaks and clumps of
flowering, golden wattle growing literally on
the sandy beaches, stretched inland for many
miles, while towards the camp we could just see
the blue Pacific showing against the high white
wall of glistening sand that separated the lake
from the sea. The heavy storm of the previous
evening had brought down a vast volume of
water, and the lagoon was quite a foot or two

higher than the sea-level. In the evening we saw the sea and lagoon on a level, and only divided by a narrow strip of sand ; another thunderstorm or a heavy sea or two would have washed this away, and a wide entrance have been formed, only to be closed up again in a few weeks. All over the placid lake (whose waters are very brackish) were parties of black swans swimming lazily to and fro, or resting asleep, and not deigning to notice the noisy ducks and waterhens around them. On the other bank long rows of pelicans stood in solemn silence. The waters were alive with fish, and, indeed, that afternoon the four of us caught whiting and bream till we were tired. One only needed to stand on the sand and fling his line into about two feet of water, when the bait would be literally rushed, for the lagoon entrance having been closed to the sea for over a month, the fish had all gathered at the sea end, their instinct telling them that it might force a passage through at any moment. After spending an hour in attempting to get a shot at a flock of ducks, we returned to camp to eat our fish and game. We lazed away two delightful days, and then, saddling our horses again, we rode northward to the little town, to the symphony of the beating surf and the rustle of the tree-clad slopes high above us to the west.

A Noble Sea Game.

J UST as my wild-eyed, touzle-headed Gilbert
Island cook brought me my early coffee
and hard ship biscuit, Toria and Vailele—
brown-skinned brother and sister—peeped in
through the window, and in their curious
bastard Samoan said 'twas a glorious morn to
fahaheke.

Now I had learned to *fahaheke* (use a surf-
board), having been instructed therein by the
youths and maidens of the village individually
and collectively. And when you have once
learned surf-swimming the game takes possession of your innermost soul like unto cycling
and golf. So I said I would come, and instantly
my young friends handed me in a surfing costume, a highly indecorous looking girdle of thin
strippings of the leaf of the pandanus palm.

147

This I blushingly declined, preferring a garment of my own design—a pair of dungaree
pants razeèd from the knees down. Then,
bidding me hurry up and meet the swimming
party on the beach, Toria and his sister ran
back to the village to attend early morning
service, to which the wooden cylinder that did
duty for a church bell was already summoning
the people.

Now, in some of the Pacific Islands surf-
swimming is one of the forbidden things, for
many of the native teachers hold the sport to
savour of the *po uli*—*i.e.*, the heathen days—
and the young folks can only indulge in the
innocent diversion away from the watchful eye
of the local Chadband and his alert myrmidons,
the village police, among whom all fines are
divided. But in this particular little island we
had for our resident missionary a young stalwart
Samoan, who did not forbid his flock to dance
or sing, nor prohibit the young girls from wearing flowers in their dark locks. And he himself was a mighty fisherman and a great diver
and swimmer, and smoked his pipe and laughed
and sang with the people out of the fulness of
his heart when they were merry, and prayed for

and consoled them in their sorrow. So we all loved Ioane, the teacher, and Eliné, his pretty young wife, and his two jolly little muddy brown infants ; for there was no other native missionary like him in all the wide Pacific.

The simple service was soon over, and then there was a great scurrying to and fro among the thatched houses, and presently in twos and threes the young people appeared, hurrying down to the beach and shouting loudly to the white man to follow. A strong breeze had sprung up during the night, and the long rolling billows, which had sped waveringly along for, perhaps, a thousand miles from beyond the western sea-rim, were sweeping now in quick succession over the wide flat stretch of reef that stood out from the northern end of the island like a huge table. Two hundred yards in width from the steep-to face it presented to the sea, it ceased, almost as abruptly as it began, in a bed of pure white sand, six feet below the surface of the water ; and this sandy bottom continued all the way from the inner edge of the reef to the line of coco-palms fringing the island beach. At low tide, when the ever-restless rollers dashed vainly against the sea-face of the reef, whose surface was then bared and shining in the sun, this long strip of sheltered water would lay quiet and undisturbed, as clear as crystal and as smooth as a sheet of glass ; but as the tide rose the waves came sweeping over the coral barrier and poured noisily over its inner ledge till the lagoon again became as surf-swept and agitated as the sea beyond. This was the favoured spot with the people for surf-swimming, for when the tide was full the surf broke heavily on the reef, and there was a clear run of half-a-mile from the starting-point on the inner face of the coral table to the soft, white beach. Besides that, there was not a single rock or mound of coral between the reef and the shore upon which a swimmer might strike—with fatal effect if the danger were not perceived in time.

The north point was quite a mile from the village, and, the tide being very high, we had to follow a path through the coconut groves instead of walking along the beach, for the swirling waves, although well spent when they reached the shore, were washing the butts of the coco-palms, whose matted roots protruded from the sand at high-water mark. In front of us

raced some scores of young children ranging
from six years of age to ten, pushing and
jostling each other in their eagerness to be first
on the scene. Although the sun was hot
already, the breeze was cool and blew strongly
in our faces when we emerged from the narrow
leafy track out upon the open strand. Then
with much shouting and laughing, and playful
thumping of brown backs and shoulders, Timi,
the master of ceremonies for the occasion, mar-
shalled us all in line and then gave the word
to go, and with a merry shout, mingled with
quavering feminine squeaks, away we sprang
into the sea, each one pushing his or her surf-
board in front, or shooting it out ahead, and
trying to reach the reef before any one else.

And now the slight regard for the conven-
tionalities that had been maintained during the
walk from the village vanished, and the fun
began—ducking and other aquatic horseplay,
hair-pulling, seizing of surf-boards and throwing them back shorewards, and wrestling
matches between the foremost swimmers. The
papalagi (white man), swimming between the
boy Toria and a short, square-built native
named Temana, had succeeded in keeping well

in the van, when he was suddenly seized by the
feet by two little imps, just as a sweeping roller
lifted him high up. And down the white man
went, and away went his surf-ooard shoreward
amid the shrieking laughs of the girls.

"Never mind," shouted Temana, shaking
his black curly head like a water-spaniel; and
seizing a board from a girl near him, and
pushing her under at the same time, he shot
it over towards me; and then Toria, with a
wrathful exclamation, caught one of the imps
who had caused my disaster and, twining his
left hand in her long, floating hair, pitched her
board away behind him. This little incident,
however, lost us our places, and amid the merry
gibes of some naked infants who were in the
ruck, we swam on in face of the slapping seas,
and at last gained the edge of the reef, which
was now alive with nude, brown-skinned figures,
trying to keep their position in the boiling
surf for the first grand " shoot " shoreward.

Between the lulls of the frequent seas the
water was only about four feet deep, and presently
some sort of order was formed, and we awaited
the next big roller. Over the outer reef it
reared its greeny crest, curled and broke with

thundering clamour, and roared its mile-line length towards us. Struggling hard to keep our feet on the slippery coral against the swift back-wash, we waited till the white wall of hissing foam was five feet away, and then flung ourselves forward flat upon our boards. Oh, how can one describe the ecstatic feeling that follows as your feet go up and your head and shoulders down, and you seem to fly through the water with the spume and froth of the mighty roller playing about your hair and hissing and singing in your ears? Half a mile away lies the beach, but you cannot see it, only the plumèd crowns of the palms swaying to and fro in the breeze ; for your head is low down, and there is nothing visible but a wavering line of shaking green. Perhaps, if you are adept enough to turn your head to right or left, you will see silhouetted against the snowy wall of foam scores and scores of black heads, and then before you can draw your breath from excitement the beach is before you, and you slip off your board as the wave that has carried you so gloriously in sweeps far up on the shore, amid the vines and creepers which enwrap the sea-laved roots of the coco-palms.

Then back again, up and down over the
seas, diving beneath any that are too high and
swift to withstand, till you reach the ledge of
the reef again and wait another chance.   Not
all together do we go this time, for now the
swimmers are widely separated, and as we swim
out we meet others coming back, flying before
the rollers under which we have to dive.   Here
and there are those who from long practice
and skill disdain to use a board ; for springing
in front of a curling sea, by a curious trick
of hollowing in the back and depressing the
head and neck, they fly in before the rolling
surge at an amazing speed, beating the water
with one hand as they go, and uttering wild
cries of triumph as they pass us, struggling
seaward.   Others there are who with both
hands held together before them, keep themselves well in position amid the boiling rush of
waters by a movement of the legs and feet alone.
But, that day, to my mind the girls looked
prettiest of all when, instead of lying prone,
they sat upon their boards, and held themselves in position by grasping the sides.   Twice,
as we swam out, did we see some twenty or
thirty of them mounted slopingly on the face

of a curling sea, and with their long, dark locks trailing behind them, rush shoreward enveloped in mist and spray like goddesses of the waves. Their shrill cries of encouragement to each other, the loud thunder of the surf as it broke upon its coral barrier, the seething hum and hiss of the roller as it impelled them to the beach, and the merry shrieks of laughter that ensued when some luckless girl over-balanced or misguided herself in the midst of the foam, lent a zest of enjoyment to the scene that made one feel himself a child again.

For two hours we swam out again and again to fly shoreward ; and at last we met together on the beach, to rest under the shade of the palms, the girls to smoke their banana-leaf *sului* of strong negro-head tobacco, and the men their pipes, while the younger boys were sent to gather us young drinking-coconuts. And then we heard a sudden cry of mingled laughter and astonishment ; for, tottering along the path, surf-board under arm, came an old man of seventy, nude to his loins.

" *Hu ! hu !* " he cried, and his wrinkled face twisted, and his toothless mouth quivered, " is old Pakia so blind and weak that he cannot *fahaheke?*   Ah, let but some of ye guide me out and set me before the surf—then will ye see."

Poor old fellow! Like an old troop-horse who dozes in a field, and whose blood tingles to some distant bugle call, the ancient, from his little hut near by, had heard our cries, and his brave old heart had awakened to the call of lusty youth. And so, earnestly begging the loan of a board from one of the swimmers, he had come to join us. And then two merry-hearted girls, taking him to the water's edge, swam out with him to the reef amid our wild cheers and laughter. They soon reached the starting-point, and then a roar of delight went up from us as we saw them place the ancient on his board, his knees to his chin, and his hands grasping the sides. Then, as a bursting roller thundered along and swept down upon them, they gave him a shove and sprang before it themselves—one on each side. And, old and half blind as he was, he came in like an arrow from the bow of a mighty archer, his scanty white locks trailing behind his poor old head like the frayed-out end of a manilla hawser, his face set, and his feeble old throat crowing a quavering, shaking note of triumph as he shot up to the very margin of the beach, amid a roar of applause from the naked and admiring spectators.

Poor old Pakia ! Well indeed art thou entitled to this stick of tobacco from the white man to console thy cheery and venerable old pagan soul in the watches of the night.

The Gigantic Albicore of Polynesia—The Takuo.

U NDER a sky of brass, and with the pitch bubbling up in every seam of our heated decks, for two days our little trading vessel had drifted to the eastward, borne steadily along by a swift, strong current. Then, just as we had lost sight of the cloud-capped peak of Ponapé, a faint line of palms stood outlined upon the quivering sea-rim, ten miles ahead. This was Ngatik, one of the many hundreds of low-lying atolls that form the greater portion of the Caroline Archipelago, in the North Pacific. As the sun sank the faint air that gave us steerage way freshened a little, and in another hour or so we had the white line of beach of the little isle in view from the deck, and knew that we should have the satisfaction by nightfall of obtaining some fresh provisions and a night's

good rest on shore. For a " furious calm,"
as our captain called it, is a horrible thing to
endure cooped up on board a small trading
vessel of seventy tons, carrying an odoriferous
cargo of copra (dried coconut), sharks' fins, and
whale oil.   Two weeks before we had lost both
our boats in the surf when struggling over
the sweeping seas on the reef at Duperrey's
Island, and our skipper thought that we could
buy at least one from the white trader living
on Ngatik ; this was our main object in
touching at such an isolated spot.

An hour before sunset we were within a mile
of the beach and saw the trader's boat being
launched and manned.   In a quarter of an hour
she came alongside, the trader jumped on
deck, and then good-naturedly offered to let
his boat's crew tow us in to an anchorage before
it became dark.   A line was soon passed into
the boat, and, aided by a light air now and
then, we went along in fine style.

Our visitor was a young, powerfully-built,
deeply-bronzed American, named Harry Stirling.
He was a great sportsman, and presently told
us that we had come to Ngatik in good time,
as the island was literally alive with pigeons—

it being the time of the year when they flocked
over from Ponapé to feed on a large red berry
which is very plentiful just after the rainy
season ; and, more than that, he said, we could
witness a great dance and feast which was to
take place on the lagoon beach that very
evening. And then, as an additional attraction,
he promised to send his boat's crew out with
me to fish for *takuo*.

Now, although I had seen the *takuo* in
Eastern Polynesia among the Tokelau and
Phœnix Groups, I had often heard old traders
in the North-West Pacific assert that the
mighty fish was absolutely unknown in the
Marshall and Caroline Archipelagoes. Doubt-
ing the correctness of this, I had several times
tried the deep water off the barrier reefs of
Strong's Island and Ponapé—for I was very
anxious to catch one of these ocean prizes—but
without success.

I soon learnt from Harry that they were
very plentiful, not only about Ngatik (outside
the reef), but in the lagoon of Providence
Island—the Arrecifos of the old Spanish navi-
gators. But the natives of Ngatik had themselves never caught a *takuo* until they were

initiated into the method of capture by Harry's boat's crew, who were natives of Pleasant Island, a lonely spot situated just south of the Equator, and between long. 165° and 170° W. These Pleasant Islanders are expert deep-sea fishermen; they are an offshoot of the Gilbert Islands people, and, although of a fierce and intractable nature, are much sought after and valued by isolated traders in Micronesia and Polynesia for their fidelity to white men, their great bodily strength, and the aversion they have to mix even with natives who are allied to them in language, customs, and mode of life generally. By the Samoans and other Eastern Polynesians the Pleasant Islanders are as much dreaded as are the warlike natives of Rubiana by the rest of the inhabitants of the murderous Solomon Group. My friend had over thirty of these people working for him on Ngatik—men, women, and children. They had followed his fortunes for some years, and, hot-tempered and quarrelsome as they were with strangers, they served him with the most unquestioning loyalty and obedience. On this occasion he had over a dozen men with him in the boat, and as they struck their broad-bladed paddles into the water they sang a weird, monotonous song in their harsh, guttural tongue, and our own crew of Rotumah men and Niué " boys " gazed at them in wondering distrust as wicked heathens —for the Pleasant Islanders would never let a missionary put foot ashore in their island home.

We dropped anchor at sunset just abreast of the trader's house, and were soon all ashore enjoying his open-handed hospitality, and surrounded by a hundred or so of the Ngatik people, who came to pay their respects to the captain and myself. They are a small, slenderly built race, and even the men looked very effeminate beside Harry Stirling's huge, brawny-backed followers.

With *takuo* running riot in my mind, I managed to evade attending the chief's great feast and dance by sending him a present and a message to the effect that " my heart was eaten up with hot desire to catch a *takuo*," and that, as the night was fine and calm, I begged his Highness to excuse me, etc., etc.

Leaving the captain and Harry, therefore, to honour the entertainment by their presence,

I leisurely set about preparing my tackle. I had a month previously bought from an American trading brig some magnificent hooks —hollow-pointed, flatted Kirby's, about 6 in. long in the shank and with a corresponding curve, and as thick as a lead pencil—big enough and strong enough for a full-grown tiger shark. My line, too, was a good one— American cotton, 32 cord, and as stout as signal halliards—just the very thing for either a *takuo* or a *pala.* Three or four of Harry's Pleasant Island natives watched me with great interest, and all expressed their admiration of my tackle, and then showed me their own —thick nine-plait coir cinnet, and heavy, barbless hooks, the points curved in so deeply as to render a barb unnecessary. These were their own manufacture, made from old fish spears.

My host's brother-in-law (Stirling had a Pleasant Island wife) was in charge of our party, which numbered six, just enough for the boat, on reaching which, two hours before dawn, we found a number of children of both sexes awaiting us, carrying baskets of cooked fowls, fish and young coconuts, which were at once

put on board ; evidently my friends intended
that we should make a night of it.

Pushing off, we paddled across the lagoon
under the most glorious starlight imaginable,
and soon reached the passage through the reef.
Here a torch of coconut leaves was lighted, and
a dozen or two of flying-fish were caught for
bait ; and then hoisting our sail, with a gentle
air from the land, we ran along the outer edge
of the reef just away from the great curling
rollers, till we rounded the weather horn of the
little island.   Lowering the sail we made all
snug, lighted pipes, and baited our hooks.   As
takuo are generally caught best when drifting,
we did not anchor, but sounded first to feel
our ground, and touched the coral bottom
with our lines at about twenty-five fathoms.
Then lowering to about twenty fathoms we
began to fish.   Each native baited with a
whole flying-fish—I used but a half of one.
For a few minutes not a sound was heard
until Tebau (Harry's brother-in-law) suddenly
darted his hand into the water and seized
a large garfish which had ventured too close.
It was a beautiful silvery scaled fish, nearly a
yard long, and just as Tebau held him up for

me to look at, saying it was better bait than
flying-fish, a young lad sitting next to me gave
a grunt, and I heard his rough cinnet line
grinding against the gunwale. In an instant
we were all on the alert. That it was a heavy
fish I could see. " Shark ? " I queried.

" No, feel the line," he replied, and the
moment I felt the jerky vibrations I knew that
it was not a shark. Presently we caught sight
of a white, wavy mass, and then up came a
large, scaleless fish called a *lahe'u.* He was
quite 20lb. or 25lb., and kicked up a tremendous row when dropped into the boat, and
accompanied his struggles by emitting a peculiar
grunting sound. Although of a bright silvery
colour, he was a most unpleasant creature to
handle, for his skin was covered with a peculiar
slimy exudation, and I was warned by my
native friends to keep my line clear, or
else I could never hold it if I got a heavy
fish on.

" We must go out further," said Tebau ;
" we are in too shallow water."

We drifted about for half an hour, and,
meeting with no luck, were just about to try
the lee side of the island, when, an hour before

sunrise, I got a run, and before I could properly feel my fish he had taken eight or ten fathoms of slack line away.

" That's a *takuo*," said Tebau.

Now did I find the advantage of a thick line in a great depth of water, and with a strong fish at the end of it ; for although only a small *takuo*—about 10lb.—this fellow gave me some trouble pulling him up, making swift dives whenever he felt the line slacken a bit, and shaking his head violently in the endeavour to free himself from the hook. As soon as I had got him in over the side one of the Pleasant Islanders lit our boat lantern, and I had a good look at my first *takuo*. In every respect but one I found him to be exactly like the common albicore in shape, colour, and markings, the only difference being the bony tail and thick laminated plates extending up from the tail for about four inches; the centre ridge of these laminations was as sharp as a knife-edge. This, however, was but a baby fish ; before daylight we caught three others, the largest of which we weighed at the trader's house, it scaled 81lb. Just as dawn began to break I hooked another, but when about half-way up he made such a fierce rush, that, seeing

I had but a fathom or two of line left inboard, I foolishly took a turn round a rowlock, and the hook snapped. Only that I was so excited and nervous, I should have remembered that I ought to have gone either forward or aft, and thus let him tow the boat a bit ; instead of this, by my remaining amidships, and the fish diving almost vertically in true albicore fashion, the boat remained stationary, and something had to go. However, I was well satisfied with our night's fishing. I had caught one of the four *takuo* taken, and my pleasure was increased when the red sun shot up from the sleeping sea and revealed the beauties of our prizes with their broad dark-blue backs, sheeny silvery sides and bellies, and bright golden fins and armour-clad tails.

From my wild, half-naked companions I learnt much of the habits of this great ocean fish, the flesh of which, despite its size when full-grown, is rather delicate even to European palates. Unlike the true albicore, which is almost a surface-swimming fish, the *takuo* haunts the coral beds at depths of from thirty to one hundred yards, or at any rate appear to do so, for they are usually caught in very

deep water, and certainly do prey upon rock-
cod, crabs, etc., that are never found in shallow
water. Then, too, they are semi-nocturnal feeders,
which is not the case of the ordinary albicore
or bonita, and at times take a bait on a moonlight night, although they bite best at dusk or
dawn, and, indeed, will occasionally follow the
baited line almost to the surface in broad daylight.
Especially is this so with a certain variety
of the species which are found in the deep
lagoon islands of the Marshall Group, such as
Mădurŏ and Arnhu Atolls ; these sometimes
attain an enormous size, scaling up to 100lb.
and 150lb. But they grow even to greater
weights than this, according to native accounts,
and I was shown the skull of one caught on
Pleasant Island in 1891, which a local trader
assured me was over 6ft. in length, " with
a body as thick as a young cow." When
hooked they make the most desperate attempts
to free the hook by a series of violent head-
shakings, after the manner of the tarpon of
Florida. Sometimes they will make a straight
dart upwards at a great speed, and then, slewing round with lightning-like rapidity, dive
almost vertically, snapping a hook or line

strong enough to hold a full-grown porpoise —the strongest and swiftest fish I know.

Six months later, when we were again cruising through this magnificent group of islands, we anchored at the lovely and fertile Kusaie (Strong's Island), the eastern outlier of the archipelago, and a resident trader there told us of a novel experience that had befallen him a week previously. He and some natives had set off in a canoe to look for turtle when they saw a huge lagoon *takuo* following them. Not having a line on board of sufficient strength to hold such a fish, one of the natives suggested that the turtle spear would do instead—they could entice him close enough to drive it into him. The line attached to the spear—a heavy piece of iron with an old-fashioned V-shaped barb — was about thirty fathoms long, and strong enough for any purpose. Some small fish were thrown overboard, and these were quickly snapped up by the *takuo*, which was very daring and hungry. In a few minutes he came quite close to the canoe, and opened his huge mouth to seize a piece of fish trailed along the surface by the white man, and at the same moment the native who held the turtle spear darted his weapon, and sent it clean through the great fish's jaw. In an instant the *takuo* "sounded," going down almost in a straight line, and the trader, fearing that line and spear would be lost, shouted to one of his crew to take a couple of turns of the line round the for'ard pole that supported the outrigger. The native—a boy— was so confused that by some mischance the end of the line, which was knotted, got jammed before he could take the necessary turns. Away went the canoe, the trader keeping it head on as well as he could by steering, for with the slightest deviation it would have either capsized or filled. Presently the fish rose a few fathoms (but still kept up a great speed), and the man for'ard attempted to drag free the knotted end of the line. But he was not quick enough, for suddenly the *takuo* went sharp about and then dived again, and in an instant over went the canoe and out went the occupants. The weight of the now filled craft seemed to drive the fish desperate, for he made tremendous struggles to free himself, rising twice to the surface and making a terrific splashing. Eventually the outrigger lashings carried away, but the for'ard outrigger pole stood, and two of the natives swam after the canoe and managed to get in. Quite a quarter of an hour, however, passed before the *takuo* was sufficiently done up to be hauled alongside and stunned by blows from a paddle. It was found that the barb of the spear had gone right through the tip of the bony upper jaw, and had scarcely injured the fish at all. Being quite impossible to take such an enormous-girthed creature into the canoe, it was towed ashore amid the plaudits of the assembled villagers. I was shown some huge strips of its sun-dried flesh by the trader, and quite believed his assertion that it weighed over three hundred pounds.

Old Samoan Days.

WE lived right merrily down there in fair Samoa, four-and-twenty years ago, in the days when our hearts were young, and those of us who had dug our trenches before the City of Fortune took no heed of the watches of the night ; for, then, to us there was no night—only long, long happy days of mirth and jollity, and the sound of women's voices from the shore, mingling with the chorus of the sailors, and the *clink, clink,* of the windlass pawls as the ships weighed anchor to sail for distant isles. And no one checked our youthful insolence of mirth ; for then there was no such thing known as the Berlin Treaty Act " for the Better Government of Samoa," with its comedy-tragedy of gorgeously bedizened Presidents, and Vice-Presidents, and Chief

172

Justices, and Lands Commissioners, and goodness knows who, of whom no one in England
would scarce have ever known, but that the
slender, wasted finger of the man who rests on
the summit of Vailima Mountain pointed at
them in bitter contempt and withering scorn, as
silly, vain people who lived in his loved Samoa.

Ah ! merry, merry times were those in the
olden days, although even then the rifles
cracked, and the bullets sang among the orange
groves along Apia beach ; for the rebel lines
were close to the town, and now and then a
basket of bleeding heads would be carried
through the town by mourning women who
beat their brown, naked breasts and made a
tagi [1] throughout the night.

There were not quite a hundred white people
living in Apia then, half of whom were Germans ; the rest were Englishmen, Americans,
and Frenchmen. But almost every day there
came a ship of some sort into the little reef-
bound harbour. Perhaps it was a big German
barque, direct from Hamburg, laden with vile
Hollands gin and cheap German trade goods ;
or a wandering, many-boated sperm whaler, with

[1] Lamentation.

storm-worn hull, putting in to refresh ere she
sailed northward and westward to the Moluccas ;
or a white-painted, blackbirding brig from the
Gilbert Islands, her armed decks crowded with
wild-eyed, brown-skinned naked savages, who
came to toil on the German plantations ; or a
Sydney trading schooner such as was ours—
long, low, and lofty sparred. Then, too, an
English or American man-of-war would look
in now and again to see how things were going,
and perhaps try some few land cases which were
brought before the captain, or make inquiries
about that Will o' the Wisp of the ocean,
Captain Bully Hayes. And the air was full of
rumours of annexation by one of the great
Powers interested in Samoa, and the Americans
mistrusted the English, and the English the
Americans, and they both hated the Germans as
much as the Samoans hated them.

    .   .   .   .   .

One day I set out to pay a visit to a native
friend—a young chief named Gafalua (Two
Fathoms). And a very good name it was, too,
for he was a man who stood over six feet on
his naked feet. He lived at a pretty little
village named Laulii, a few miles northward

from Apia, and I had to cross several tiny rivers ere I came to the final stretch of beach that led to the place. The air was full of a sweet summer softness, and as I walked along the firm, hard sand, with the cool shade of the forest on my right, and the wide sweep of reef-bound water on my left, I felt a strange but delightful elasticity of spirits. Now and then a native carrying a basket of fruit or vegetables would pass me with swinging tread, and give me a kindly *Talofa!* [1] or, perhaps, setting down his load, would stop and chat for a few minutes.

Presently, as I turned into a bend of the beach, I came across a party of some eight or ten people, seated under the shade of a coconut tree, and talking eagerly together. Most of them were old acquaintances, so laying down my gun, I acepted their invitation to stay and *nofo ma tala tala fua*, i.e., rest and indulge in a little talk ; " for," said one of them, " we have news. There is now an American man-of-war at anchor in Saluafata. She came there last night, and now are we moved in our minds to know what this may mean to Samoa. What do *you* think ? "

---

[1] The Samoan salutation—" My love to you."

I shook my head. How could I tell? I knew nothing of these things. "Perhaps," I suggested, "she has but come into Saluafata Harbour to give the men liberty, for there is much sickness in Apia."

"Aye," said one man, with a sigh, "'tis like enough. But are we never to know whether America or England will put their hand over us, or are we all to be swallowed up by the Germans?"

To this I could say nothing, only sympathise; and then I learned to my great pleasure that the man-of-war was a ship I knew, and her doctor was an old friend of mine whose acquaintance I had made in the Caroline Islands a year or two before, when I was making my first voyage as supercargo. So after smoking a cigarette with my friends I bade them goodbye, and set out again for Laulii.

An hour later I reached the village, and was warmly greeted by some forty or fifty people of both sexes. Gafalua, they told me, had gone to Saluafata to visit the warship, yet if I would but send a message to him he would quickly hasten back to greet his white friend. And as they clustered around me, each one volunteering to be messenger, the chief's daughter, Vaitupu, a charming girl of fourteen, accompanied by a younger brother, ran up and embraced me with the greatest demonstrations of joy, for I was once an old comrade of theirs in days gone by in many a fishing trip and forest ramble along the shores of both Upolu and Savaii—the two principal islands of the group. And then, having sent off a message to Gafalua and written a note on the leaf of my pocket-book to the doctor of the warship, I resigned myself to the never-ending attentions of my native friends. By and by, after I had eaten some baked fish and drank a young coconut, the whole of the elder women in the village entered the house, and seating themselves in a semicircle before me, plied me with questions as to where I had been all these long, long moons. Had I seen the black people of the Solomon Islands—they who ate men? Was it true, the tale they had heard of a trading ship coming from America to sell the people repeating rifles on long credit? Had I seen the great circus in Nui Silani (New Zealand) of which Pili had told them—a circus in which one man jumped over four-and-twenty horses? Or was Pili only a liar?

13

And then one old dame bent forward, and put *the* question :

"Why does the American man-of-war come here ?   Has she come to help our king Malietoa to fight the rebels and drive away the Germans ? "

I could only say that I could not tell ; for I had been away from Samoa for more than a year.

The ancient lady rolls herself a cigarette in a meditative manner, and then looks gloomily out before her upon the sea-front.

"*Tah !* " she says at last.   " It is always the same, always, always.   'Tis all talk, talk, talk. One day it is, ' Ah, next moon American and English soldiers will come, and they will set up Malietoa, and the flash of their bayonets shall blind his cruel enemies, so that they will shake and turn pale ! ' ; or, ' Not next moon, but the one after, a big man-of-war will come from Peretania, and bring hundreds of red-coated fighting men, whose chief will draw a line with his sword on the beach at Matafele, and say to the Germans, ' Keep thee all there, beyond that line, and within thine own bought land ; step over but a hand's space and thou shalt hear

the rattle of a thousand English guns." But
they never come—only the men-of-war, whose
captains say to our chiefs, ' Not this time ; but
by and by we shall help thee." And then at
night time they make their ships bright with
many lights, and the *tamaitai papalagi* (white
ladies) from Apia and Matafele put on beautiful
dresses, and they all dance and sing and laugh,
and think no more of us Samoans ; and in the
morning, or in a day, or two days, the ships go
away, and we Samoans are like fools, and hang
our heads. Then the Consuls say, ' Hush ! be
wise and wait ' ; but the Consuls are liars ; one
gives us fair words and sweet smiles and says,
' Vitolia (Victoria) is great, she loves you
Samoans, and will help you ; but you must
not want to fight the great German nation.
You must come to us, and we shall send a
letter to the great chiefs in Peretania (England) and America, and—*by and by help will
come.*'"

It is impossible to describe the sneering, bitter
emphasis the old woman gave to her last half-
a-dozen words, imitating, as she did to perfection, the voice of the then British Consul. That
gentleman is long since dead ; and whilst his

genial social qualities will long be remembered by those who knew him, his foolish official acts made him many enemies, and caused intense bitterness of feeling among the natives.

The grey-haired dame smoked on in silence, and then a tall, lithe-limbed girl rose from beside the old woman, and came over to me, and, taking the inevitable cigarette from her lips, offered it to me. "She is my son's wife," explains the old woman, as the pretty creature seats herself again ; and then this soft-eyed, sweet-voiced girl says, with an innocent, childish laugh : "*Tah!* I love to hear the *pa fana* (firing of guns). My husband took two heads at Mulinu'u once. Fighting will come by and by, my mother, and your son shall bring you ' red bread-fruit ' to look at again."

For the edification of my readers I may explain that the term "red bread-fruit" was then the Samoan slang for decapitated heads. This amiable young lady evidently had a full share of the Samoan women's spirit that causes them to very often leave the care of their children and houses to the very oldest of their sex, and follow the fortunes of their husbands or lovers to the camp.

Another hour passed, and then there came a rush of excited children along the narrow, shady path that led into the village from the northward. " Gafalua 'is coming," they cried pantingly, " and with him there are two officers from the ship — a little, dark-faced man with a black moustache, and a big, fat man."

I ran out to meet them, and in a few minutes was shaking hands both with Dr. T—— from the warship, and my native friend, the chief. The doctor, who was in uniform, was bound for Apia, in company with Lieutenant D——, to make inquiries concerning the outbreak of sickness there, the commander of the corvette not liking to take the ship to Apia until he had satisfied himself that there was no risk in so doing. The doctor agreed to meet me in Apia on the following day, and, if possible, join Gafalua and myself in a mountain excursion to the other side of the island, where we were to remain for a couple of days at the village of Safata.

.   .   .   .   .

Early next morning, accompanied by Gafalua, his son and daughter, and four or five young men and women carrying cooked food for the journey to Safata, I set out for Apia—a three hours' walk. Inquiring for the doctor at the American Consulate, I found a note for me saying that he and Lieutenant D—— had returned to the ship to report that Apia was too unhealthy just then for her to make a stay at ; also that he would apply for a few days' leave, and expected to return in the evening. Leaving Gafalua and his followers at the native village of Matautu, I returned to my own ship, and gathered together a few extra traps for the journey. As I was pulled ashore the boat had to pass under the stern of a large Sydney trading brig, whose captain hailed me, and asked me to come aboard. With him were his wife and daughter, who were making a visit to Samoa after an absence of some years, and, curiously enough, that very morning they had been discussing means of going to Safata to spend a few days there with the resident missionary, who was an old friend. To sail round in a cutter would mean at least a two days' voyage and a vast amount of discomfort.

" Why not come with us ? " I suggested. " We can leave this evening, sleep at Magiagi " (a village a few miles from Apia), " start early

in the morning, and be at Safata in the afternoon."

In five minutes everything was arranged ; the
two ladies were to meet the rest of the party
at the Vaisigago ferry, and I hurried ashore,
delighted at my luck in securing such charming
companions. Both Mrs. Hollister and her
daughter spoke Samoan, and were great
favourites with the natives of the Apia
district. Early in the afternoon the doctor
returned—happy with four days' leave—and
we were at once joined by Gafalua and his
people. At the ferry we found the two white
ladies awaiting us, with another addition to our
party—the half-caste wife of an American store-
keeper. Then we started.

Our way lay along the principal roadway or
street of Apia, as far as the white-walled native
church, and then made a détour to the left,
inland. The town of Apia, or properly speaking, the towns of Apia and Matafele combined,
are laid out in a very irregular manner ; and the
main street follows the curves of the beach.

The sun was somewhat fierce, and we hailed
with delight the cool, shaded road which lay
before us after we turned off from the town.

It had been raining a few days previously, and the middle of the road was somewhat muddy, but the side-paths were fairly dry for the ladies, who declined the offer of our natives to be carried till we reached the first resting-place. The soil here was a rich, red loam ; and from the beach for nearly two miles inland the road lay through banana and taro plantations, with here and there small villages inhabited by the adherents of Malietoa. Every now and then natives would pass us—generally women— with loads of taro, yams, or fruit ; and it was pleasant to note the courteous manner in which they left the dry side-walk and stood in the boggy centre of the road while we passed. By nearly every one we were greeted with a smile and offer of fruit for the ladies, or a coconut to drink.

About two hours after crossing the Vaisigago and proceeding in a south-easterly direction, we heard the sound of a cataract, and presently we again got a sight of the river through the trees. We turned off at this spot to look at the favourite bathing place of the white residents, a deep pool of some fifty yards in length, surrounded by a thick, tropical vegetation. The Vaisigago here was a noisy, brawling little stream, and at the head of the pool was a gorge, between the black, gloomy sides of which the bright, clear water came rushing down with many a swirl and hiss, and forming in a deep, rocky depression a miniature lake. Our carriers laid down their burdens, and waited whilst we sat on the edge of the pool, to enjoy for a few minutes, the pretty sight. The water was full of fish resembling English trout ; and there were also two or three kinds of a small size, and precisely similar to those found in the rivers of Northern Australia. One of the natives went down into the creek, where the water was shallow, and groping with his hands under the boulders, caught two or three large shrimps : great fat, brown fellows, that jumped about in a most active manner when laid on the rocks beside us, making a peculiar snapping noise with their huge nippers. Taking one up, the native bit its head off, and then breaking the body into three or four pieces, desired Miss Hollister to throw them into the water. The instant the dismembered fragments touched the surface there was a rush of fish, and the glassy surface of the pool—for in the centre

there was no apparent current—was swirled and
splashed and eddied about. The doctor was so
excited at such promising indications of sport
that he announced his intention of returning to
Apia, and borrowing a rod and tackle ; but we
promised him that on our return we should pay
another visit to the pool, and make a day of it.
This spot is locally known as " Hamilton's
Pool," being named after the then port pilot,
Captain Edward Hamilton. Many years ago,
when H.M.S. *Pearl* was in Samoa, that ill-fated
and gallant sailor, Commodore Goodenough,
who was fated to die by the poisoned arrows
of the savages of the Santa Cruz Group,
delighted to make his way here and drink in
the romantic beauty of the scene.

But we could not linger. We had still some
miles to travel ere we reached the bush village
where we were to rest for the night. Shouldering their burdens, our carriers move briskly
along, and presently we notice that we have
almost reached the border of the narrow belt of
littoral that lies at the back of Apia ; for the
road now presents a gradual but very decided
ascent. Every now and then we hear the deep
booming note of the wild pigeons, and slip cart-

ridges into our guns in readiness for a chance shot, as even at this short distance from the town the great, blue-plumaged birds are to be met with. The road has become narrower, and in the place of the tall, slender coco-palms, growing so thickly in the flat country, we see all round us the great *masoi* and *tamanu* trees, towering up high above all their fellows of the wood. We meet very few natives now, and pass no more plantations. Every now and then the *fuia*, the Polynesian blackbird, utters his shrill, sharp note, and flitting in front of us perches on an overhanging branch, leaning his head on one side in a pert, impudent manner, and saucily staring with his beady black eye at the intruders. Bird life is plentiful here. Flocks of gay, bright little paroquets dart in quick flashes of colour among the undergrowth of the forest ; while overhead there fills the air the soft cooing of thousands of ring-doves. Well have the Samoans named the ring-dove *manu-tagi*—the bird that " cries "—for there is to their imaginative natures an undercurrent of sadness in the gentle cooing notes that fill the silent mountain forest with their plaintive melody, and which is rendered the more marked by the shrill scream of the paroquets, and the proud, haughty " boom ! " of the red-crested pigeon.

Now we near the village and the deep quiet of the forest is broken by sounds like chopping and tapping on wood. It is the native women, beating out with heavy wooden mallets the bark of the paper mulberry to make *tappa*, the native cloth. Our natives quicken their steps and break into song ; the sounds from the village cease, and then we hear plainly enough the soft voices of the women borne through the forest in an answering chorus of welcome. Ten minutes more, the ladies stepping out bravely in our midst, and we round the bend of the track, and there before us is a pretty little Arcadian-Polynesian village of some ten or a dozen thatch-covered houses. In the centre stands the largest edifice, a great mushroom-roofed house, open at the sides, and the floor covered with rough but clean mats made from the coconut leaf. Seated in the house are some five or six women, engaged in making *tappa* ; but they hastily lay their implements aside, and one, quite an ancient lady, bids us come in ; and, as is ever the case in Samoa with European or American travellers, welcomes us. We all file in, and in default of chairs or stools sit with our backs against the supporting posts of the house, whilst the women reach down from cross-beams overhead huge bundles of soft white mats with gaily-ornamented edges, and spread them in the centre of the house. So far, the old woman alone has spoken, it being considered the height of bad breeding by Samoans for any one to speak to or question strangers in public, until the chief or chieftainess in authority has done so. The mats being spread out, and having taken our seats cross-legged thereon, Samoan style, the old dame, in a slow, set speech, gave us her name, and said that her grandson, the chief of the village, with all his fighting men, were away at a *Fono* or native political meeting, and would not return till night, winding up her remarks by regretting that we had sent no notice of our coming, so that food and houses might be made ready for us ; but that if our " young men " would assist she would have a pig killed and get food ready instantly.

No sooner said than done ! Up jumps Talamai, one of our carriers, and disappears at

the rear of the houses ; and then arises a horrible
squealing, and much laughter from the women
and girls, as a small black porker is dragged
before the dame to inspect.   She gives a nod.
Thump ! a blow from a heavy club terminates
the animal's woes, and the carcase is dragged
off by our carriers and the women, many more
of whom are now present, having come in from
the plantations with vegetables and fruit.

Then how the native girls cluster round our
two fair fellow-travellers, and press fruit and
young coconuts upon them ; already they have
made a couch of layers of *tappa*, with a soft roll
of finely-worked mats for a pillow, and the two
white ladies recline thereon and look happy, and
talk away in Samoan to the girls.

So we smoke and chat till a wild-eyed urchin
calls out to the women, and announces that the
meal is ready to be taken from the oven of leaves
and stones.   Away run our hostesses, and in
five minutes return with roasted pork, fish, taro
and baked plantains, which are laid out on
platters made of interwoven coconut leaves.   In
the centre is placed a great pile of green coconuts.   The two ladies are served with food on
their couch ; but the doctor and myself seat

ourselves cross-legged on the ground and eat in
thorough native fashion. Our entertainers sit
each one behind a guest, and with a *fue* (or fly-
flap) brush away the flies. Never a word is
spoken by any of them except in a whisper ;
the young unmarried girls devote themselves to
Mrs. and Miss Hollister, and leave us to be
waited upon by the older women. This is intended as a special mark of respect to us ;
for to receive attention and consideration from
elderly people in Samoa is looked upon as a
graceful compliment.

Our meal finished, we fill and light our
pipes, and " lay around loose," as the doctor
calls it, to watch the first shadows of sunset
close round the little village. Darkness comes
on very quickly in these latitudes ; and soon
from every house the evening fires send fitful
flashes of light through their interwoven sides.
The wild-eyed, Italian-looking boy takes a *tappa*
mallet and strikes a long wooden cylinder
standing out in the gravelled village square. It
is the signal for evening prayer ; and then, ere
the rolling echoes of this primitive substitute for
a church bell have ceased to reverberate adown
the gloom-enshrouded forest, the women and

children gather in the house, and decorously
seat themselves round the sides.   One of our
carriers is the young Amazonian who made the
pleasant remark anent the " red bread-fruit " at
Gafalua's village.   She looks at her *Pese Viiga*
(hymn-book), and says " *Pese lua sefulu* "—
hymn 20—and then her clear bird-like notes
lead the singing.

" Well, they certainly *can* sing," says the
doctor, as the melodious voices of the women
blend with the deeper tones of our stalwart carriers in a translation of " The Living Fountain."
The singing ceases, and then one of the carriers,
a big, burly, black-bearded fellow, bends his head
and utters a short prayer.   The demeanour of
these simple natives was a revelation to the
doctor ; and at the conclusion of the short
service he asked them to show him some of their
books.   They brought him great, heavily-bound
translations of the Old and New Testament,
hymn-books, and others of a devotional character ; published in London by the British and
Foreign Bible Society ; and, indeed, the doctor
admitted that the knowledge displayed by some
of the women made him " feel rather down
in Scripture history."

Presently from out the darkness of the forest depths sounds the murmuring of voices. It is the men of the village returning from the *Fono*. Nearer and nearer they come, and now the women make the fires blaze up brightly by throwing on them the shells of coconuts. Here are the men—twenty of them—and a brave sight they make, as with a steady tramp, they march two deep over the gravelly square, the firelight playing fitfully on their oil-glistening, copper-coloured bodies, and shouldered rifles. Every man is in full fighting fig—bodies oiled, hair tied up over the crown of the head with a narrow band of Turkey red cloth, and round their waists broad leather belts with cartridge pouches. Some carry those long, ugly, but business-like looking implements, the *Nifa-oti*, or death knife, used expressly for decapitation. A few have heavy revolvers of a superior pattern, and tied round the brawny arms and legs of all are ornaments of white shells or green and scarlet leaves intermingled. The chief calls halt, and then in a semi-military fashion dismisses them, and each seeks his house, their women-kind following.

Stooping his tall frame, the chief enters the

14

big house, and in a quiet, dignified manner
shakes hands with his visitors, and acknowledges
former acquaintance with me by holding my
hand and patting gently on the back of it—a
custom that is followed in some parts of Polynesia, denoting pleasure at meeting a friend.
He does not shake hands with Mrs. Hollister
and her daughter, but, like a well-bred Samoan, -
sits himself cross-legged in front of them a few
paces distant, and, lowering his eyes, gives the
proper Samoan greeting to women of position,
Ua'e afio mai, tamaitai, which rendered in English is, "Your highnesses have come." His
mother brings him food, and then we sit round
and smoke in silence whilst the doctor fumbles
about our traps and produces a couple of bottles
and glasses, and uncorking one asks the chief to
"take a taste." His grandmother frowns dis-
approval as he pours out a "nip" that would
please a second mate, and then, the big man,
looking at us with a smile, says, *To fa, tamaitai
ma alii* (good-night, ladies and gentlemen), rolls
himself in his white *tappa* covering, and placing
his head on a curiously-shaped bamboo pillow,
is soon asleep. Simultaneously we follow suit.
The ladies, in accordance with a Samoan custom,

retire to sleep in a separate house inhabited by the *Ana luma,* or unmarried women, who escort them thither by the light of a torch.

II.

We were awakened at sunrise by the villagers, and whilst the three ladies were making their toilettes, the doctor and I, accompanied by Gafalua and the chief of the village, went to bathe in the mountain stream near by. This was a feeder of the Vaisigago, and, like that stream, its waters were of a surpassing clearness, and full of small fish and prawns. Returning to the village we found our breakfast awaiting us, and everything in readiness for a start. Half an hour later we set out, escorted for the first six or eight miles by the young women and children of the village, who insisted upon relieving our carriers of their burdens. About noon we reached the summit of the mountain range which traverses Upolu from east to west, and here we rested awhile before beginning the descent to the southern shore, and to say farewell to our companions from Magiagi, many of whom wished to accompany us to Safata ; but on account of there being ill-blood

between the two places they dared not. Only a
few months before, so they told us, a war party
of Safatans had made an attack on their village,
but had been beaten off ; some heads were taken
on both sides, and the Safatans had retreated,
vowing vengeance.

After lunch, which we ate under a huge
banyan tree, we began our march again, and in
a few minutes emerged from the gloom of the
mountain forest out upon the verge of a plateau
overlooking the coast for a dozen miles east and
west.   But much as we desired to stay awhile
and feast ourselves upon the gorgeous panorama
of tropical beauty that lay beneath us, we could
not, for there were dark clouds sweeping up
from the north, and a deluge of rain might fall
upon us at any moment.   So off we started
down the steep and slippery path, catching hold
of vines, hanging creepers, and branches of
trees, to save ourselves from getting to the base
of the mountain too quickly.   Gafalua had sent
Vaitupu and her brother on to announce the
approach of a *malaga* (a party of visitors), and
soon after we reached the level ground, and just
as the first drops of rain began to fall, we heard
the sounds of a native drum beating—the people

were being summoned together to make preparations. Soon we gained the outskirts of Safata,
and from every house we received invitations
to enter and rest till the rain ceased, but we
pressed on, and a quarter of an hour later
entered the village itself, where we were warmly
welcomed by the chief of the place. The three
ladies found the missionary and his wife awaiting
them, and promising to call upon them at the
mission-house on the following day, the doctor
and I bade them goodbye, and took up our
quarters with Gafalua and his two children in a
house specially set apart for us. A bowl of
kava was being prepared, and this we drank
with our entertainers, and then prepared to
make ourselves comfortable for the night. As
the mosquitoes were bad, our host had rigged up
a screen of fine muslin for each of the white men,
a large one for Gafalua and his children, and
many smaller ones for the rest of our company.
During the night the rain fell in torrents, but
we heeded it not, for we were tired out with
our twenty miles' walk, and the natives per-
ceiving our fatigue left us to ourselves at an
early hour, after arranging a shooting and
fishing excursion on the following morning.

A lovely sunrise greeted us when we awoke, and after eating a hurried breakfast of roast fowl and taro, we started, accompanied by Gafalua, his two children, and one or two Safata natives. We were to fish along the edges of the reef at a spot where it formed a miniature lagoon, and where, we were assured by Vaitupu, who knew the place well, we should have plenty of sport.

The sweet-scented *masoi* and cedar trees that fringed the forest, extending from the foot of the mountain to the beach, gave shelter from the rays of the sun to hundreds of the great blue-plumaged, scarlet-crested pigeons, and our progress was somewhat retarded in picking up the prizes that fell to the doctor's breech-loader. Within an hour or so of leaving the village we had secured enough to satisfy us all, and the boy fairly staggered under a load of fat, juicy birds. On reaching the beach we found a small native house, built under a giant bread-fruit tree, and untenanted. Into this we bundled our belongings, and set about rigging up our fishing-tackle. The doctor, taking his cue from me, elected to fish with a hand line, looking aghast at the gigantic proportions of the rod offered to him by Gafalua, who, in his turn, gazed with astonishment at the doctor as he noticed him tying a large steel "Kirby" to the end of his line. "No good," says Gafalua, "fish Samoa no like black hook; Samoa fish-hook very good," and displaying to the doctor a large mother-of-pearl fish hook, a marvel of ingenuity and strength. However, the doctor thought his way best, and so off we go. My young friends had not forgotten to bring me a pair of native sandals, woven from the tough fibre of the coconut, and once much used in Samoa; so, discarding my boots, I tied on the sandals in the orthodox manner, eliciting from the natives the laudatory exclamation, *Si tagata Samoa, lava*— "Like a son of the soil."

Then, all being in readiness, we start for the reef.

.   .   .   .   .

The deep, calm waters of the pool are protected on three sides by the coral reef; on the seaward side there is a narrow passage, just wide enough for a small craft to sail through, and through this the spent billows of the Pacific roll lazily and sink to rest in the quiet depths of the lagoon waters. We make our way over

the dry coral (for it is low tide), and take up
our positions where we can drop our lines
directly beneath us into the water.

The doctor stands on a little knoll of coral
nearest the beach. Gafalua, his son and daughter
and myself go further out towards the outer
reef, and we are just about to drop our lines
when a cry of alarm from the doctor is followed
by a shriek of laughter from the girl, as a huge,
yellow eel, with red eyes and snaky head, raises
its sinuous body from out its coral niche beneath
the surgeon's feet, and shows its glistening,
needle-like fangs. The doctor seizes a piece
of coral and strikes it a stunning blow on the
head, and his attendant native gives the hideous
sea-serpent the *coup de grace* by snicking off its
head with his long knife. Tough customers,
these eels ; minus his head he still wriggles and
twists his greasy, orange-yellow body about, as
if losing his head were a matter of no particular
moment. The doctor baits his hook with a small
bit of fish and throws out his line. Gafalua,
poising himself on a little coral knoll, lowers
his rod and trails the shining pearl-shell hook,
innocent of bait, backwards and forwards
through the water, and then Vaitupu calls out

triumphantly, " *Aue!* my father is first," and
sure enough the stout pole in the chief's hand
is bending and straining under the weight of a
heavy fish. What a splashing and froth he
makes as he comes to the surface, and then with
a dexterous swing Gafalua lands a magnificent
blue and yellow groper—weight about 10 lbs.
Beside us now stands Vaitupu, gaff in hand,
her dark eyes dancing with excitement, for the
doctor has wagered me a dollar he lands a fish
before I do.

"Here, here, O, my dear friend," cries
Vaitupu, "drop your line here ; down there
in that deep blue valley between the rocks are
the great big *gatala* (rock cod). Oh, such
fish, as big as a shark."

Baiting with a small, wonderfully coloured
fish, I drop my line into the " blue valley,"
while the girl and I watch the bait sinking
slowly, slowly down, till it is almost lost to
sight. A dark, misty shape rises up from
the depths below, and Vaitupu clutches my
arm.

"*Aue!* it is a *gatala;* strike, strike, my
friend."

No need for that, Vaitupu ; a sharp tug at

the line nearly capsizes me, and *gatala* makes a bolt. My tackle is, as the doctor says, thick enough to throw a buffalo, so no fear on that score ; and now, with a soft chuckle of delight, the girl lends her aid, and we pull up hand over hand.

"*Aue!*" says the little maid ; "surely it is the king of all *gatala*, it is so heavy."

Whiz ! and away he goes again, nearly taking the line away from us ; gently now, he's turned · again, and we haul up quickly. Ah ! there he is in sight now ; a great mottled-scaled fish with gleams of gold along his broad, noble back. "Good boy," calls out the doctor, "stick to him," and the two natives give a loud *Aue!* of satisfaction as the fish comes to the surface, struggling and splashing like a young alligator. Bravely done, Vaitupu ! She stoops over the coral ledge, thrusts her right hand under his great gaping gills, and planting herself in a sitting posture, hangs on right bravely, although the great strength of the fish nearly drags her over the reef. Leaping from knoll to knoll over the distance that separates us, Gafalua comes to our aid, and then reaching down his great brawny, brown hand, he too seizes *gatala* under the gills, lifts him clear, and tosses him, lashing and struggling savagely, on the reef. *Io triumphe;* or rather *Aue!* We have conquered ; and the blushing, panting Vaitupu smiles appreciation to the doctor's encomiums of her pluck.

"Hurrah!" exclaims the medico, as he grasps the slippery prize with both hands by the tail, and attempts to lift it up. "What a pity we can't take him back to Apia with us, and see him served up on the corvette's table. I guess he weighs forty pounds, too, or more."

We take up our positions again, and now both Gafalua and the other natives land fish fast enough ; mostly a species of trumpeter about 4lb. or 6lb. weight. The doctor gazes sadly at his line, not a sign of a bite yet, and turns for solace to his cigar case, when he starts up and gives an excited jerk at his line. "Hurrah! got one this time," he calls out, and some ten fathoms away, near the surface of the water, we see the silvery sheen of a long slender fish like an attenuated salmon. An eccentric fellow this, for instead of allowing himself to be pulled in like any well-regulated member of his tribe, he executes some astonishing gymnastic feats, jumping clear out of the

water and coming down again with a sounding
thwack, then darting with lightning speed to
port and then to starboard ; and, as he realises
it is not a joke, making a wild dive deep down
into the coral caverns of the lagoon. But
the doctor keeps a steady pull, and with a
cry of triumph he lands his fish at last
upon the rocks.

"*Aue !* " shouts the lively Vaitupu. " Oh,
Misi Fo Ma'i (medicine man) ; oh, clever
American, you, too, are a lucky man thus to
catch such a fish with a steel hook."

And now the calm waters of the pool begin
to swell, and gently lap the sides of the coral
rocks ; it is the tide turning, and the place
seems alive with fish of all sorts of colours and
shapes. Quickly as we drop our lines, there is
a tug and a splash, and every one of our party
is too actively employed on his own account to
heed the prowess of his companions. Half an
hour or more and we retire from the field of our
exploits to the little house on the beach, and,
whilst resting on the mats, have the pleasurable
satisfaction of seeing the boy laboriously
dragging our captures over the coral reef and
depositing them on the beach.

Grateful enough it is to rest after our labours and eat the cold pigeon and taro and bread-fruit, which the nimble fingers of Vaitupu spreads out on extemporised platters of coconut leaves. She is now at home with the doctor, and laughs gaily as she sees him endeavouring to open a young coconut. Her *tiputa* is thrown aside over one shoulder, revealing all the budding beauty of coming womanhood, and round her head she has already entwined a wreath of scarlet hibiscus flowers, gathered from a bush that flaunts its wealth of flowers and foliage near by.

" Father," she says with a laugh to the giant Samoan, " let us wait here till it is cool, and this clever gentleman from the American fighting ship will tell us *tala* about many things ; of the great guns that load in the ' belly ' ; of the iron devil-fish (torpedoes) that swim under the sea, and go under the bottoms of ships and *bite* them, and then blow them up, so that all the men are drowned ; and ask him if he has ever killed any people ; and if he has a wife in America, and is she young and pretty ; and could the American fighting ship sink the big German man-of-war that was at Apia last year,

whose captain had red whiskers and a pot
belly?"

And there we lay and smoked and talked,
and gazed sleepily out upon the sparkling sea
with the long line of foaming, reef-bound surf
far below us, till the first air of the land breeze
crept down to us from the mountains.   Then,
shouldering our burdens, we returned to the
village to watch an evening dance, and then
sleep peacefully till the morn.

  .   .   .   .   .

The morning for our return to Apia broke
brightly, with the booming of the feeding
pigeons, and the shrill cries of the gaily-hued
paroquets, as they flitted from bough to bough
in the bread-fruit grove surrounding the town.
The doctor had been up and away as the first
streak of sunrise pierced through the lattice-
worked sides of the house, to walk to the
mission house and bid farewell to the ladies,
who had sent us word that they had decided
to stay at Safata for a week.   As Gafalua and
myself were having our breakfast, we saw him
striding down the leafy path in company with
the missionary, who had returned with him to
say goodbye.   We made quite a strong party

going back, as, although we left the ladies behind at the mission, we found awaiting us some twenty natives of both sexes, who begged to be allowed to join our party, as they had business in Apia. The more the merrier, we say ; and as we have already said farewell to the old chief and the principal people of the town, we now rub noses with the chief ladies thereof, and depart amidst a chorus of good wishes.

But I must not forget. It was Gafalua's intention to leave Vaitupu with some of her Safata relatives for a few weeks, and with tears of vexation dimming her eyes she had said farewell to us at the village. The girl had quite won our hearts by her amiable and pleasing manners, and so the doctor and I, joining forces, begged her father to let the " little maid " cross the island again, and see the fighting ship with its guns that " loaded from behind."

" Only let me go with you," she pleaded, " and I shall be as silent as the dead. When we get to Apia, is not my cousin, Manumea, there ? And I can stay there with her while you, my father, go to the *olo* (forts) of the Tua

Masaga.   But I, oh, most of all, I want to see
the big man-of-war."

The burly chief looked at his daughter, and
then at myself and the doctor, and turning to
the girl, patted her hand affectionately.   "Thou
shall come, little one," he said at last, with a
smile.

We followed the same road that had brought
us to Safata, and as we struck deeper into the
leaf-covered arcades of the forest, we lost the
low murmuring of the breakers as they dashed
upon the outer barrier reef, and heard the
sudden calls of the pigeons resounding and
echoing all around us.   The morning dew
was still heavy upon the trees, and as the
birds flew away from or alighted upon them,
a shower of pearly drops fell to the ground ;
then ever and anon we heard the shrill, cackling
note of the wild cock, as with outspread wings
and scurrying feet he fled before us to his
hiding-place in some vine-clad covert.   Two
miles more, and we had crossed the narrow
belt of littoral, and were ascending the mountain
path, and now the vegetation grew denser at
every step ; for the sides of the mountain were
clothed with a verdant jungle through which

the rays even of the mid-day sun could scarcely penetrate. The path was, however, well worn, although in some places very slippery and precipitous. We envied the ease with which our native friends made the ascent, whilst we, with our boots clogged with the tough, adhesive red clay, every now and then slipped and fell.

An hour before noon we had reached the summit of the range, and with a sigh of relief assented to Gafalua's suggestion to rest for an hour or so. And so we leant our weary backs against the buttressed trunk of a great white-barked tree, and enjoyed to our full the beautiful scene below.

The trade wind was very fresh, and had tipped with " white horses " the blue bosom of the Pacific ; but away to the southward, where the outer reef reared its solid barrier against the ocean roll, there showed within its long sweeping curve the green, placid waters of shallow depth that glinted and sparkled in the tropic sun, and about the distant rush and roar of the breakers as they fell upon the reef ascended a misty haze that hovered and wavered perpetually above the swirling sheets of foam sweeping across the coral rock. Sometimes,

15

when the waving branches above our heads
ceased their soughing for a moment or two,
we heard from seaward a faint murmuring
sound that we knew was the voice of the ocean
borne to us on the breeze.　Far down below us
we saw through an opening in the forest the
thatched houses of the village, and our thoughts
went back to the kindly, honest-hearted people
who dwelt there.　To the northward of us was
hilly, undulating country, and from the sides of
the lesser hills we saw clouds of smoke ascending, showing that the men of the bush villages
were at work clearing their yam plantations.
It was a scene like to many such that may be
viewed almost anywhere in the high mountainous isles of the Pacific, but to us at that
moment it seemed the very perfection of tropic
loveliness.

.　　　.　　　.　　　.　　　.

We reached Apia as darkness fell ; and then,
bidding goodbye to the doctor and Gafalua and
the little maid, I hurried aboard our schooner,
and found that she was only awaiting my return
to sail at daylight.

And as the red sun shot up from the sea,
the sharp bows of our little vessel cleft the

swelling blue as she stood away northward and westward toward the distant Carolines, and long before noon Upolu was but a misty outline astern.

The King's Artillerymen.

THE story of the cutting-off of the London privateer *Port-au-Prince* by the natives of Lifuka in the Friendly Islands, in 1805, is pretty well known to students of the earlier history of New South Wales, for a full report of the massacre figures among the official records of the settlement. The *Port-au-Prince*, it may be mentioned, had in the earlier part of her voyage along the coast of South America captured a number of Spanish prizes, two of which had been despatched with prize crews to Port Jackson, where they arrived safely, and were duly sold a few months after the former messmates of those that brought them there had been savagely slaughtered in the Friendlies. The events of the *Port-au-Prince's* remarkable voyage were subsequently made known by the

publication in London, in 1810, of Mr. William
Mariner's " Tonga Islands." This Mariner was
a youthful friend of Captain Duck, the master
of the privateer, and seems to have been spared
by the natives on account of the friendly feeling
entertained for him by the leading chief—Finau
—and a lesser chief named Vaka-ta-bula. Some
others of the crew who happened to be on shore
at the time of the massacre of the rest of the
Port-au-Prince's company were also spared, and
these men, together with young Mariner, were
afterwards employed by Finau in aiding him to
conquer the people of Tongatabu, the main
island of the Friendly Group ; and it is from
Mr. Mariner's graphic narrative of his five
years' sojourn in the islands that the following
particulars are taken.

A few weeks after the destruction of the
greater number of the unfortunate crew of the
privateer, Finau intimated to the survivors his
intention of conquering Tongatabu, with whose
people he was at variance, and that, as the
carronades of the privateer, with plenty of
ammunition, had been saved, he wished the
English sailors to take charge of the guns, and
serve them in the reduction of the principal

fortress at Nukualofa, the capital of Tongatabu.
With that object in view, Finau, accompanied
by the pick of his warriors and Mariner and
fifteen other Englishmen, sailed from Lifuka,
in the Haapai Group, in a number of large war
canoes.

    .   .   .   .   .

The site of that fortress is still visible, and
a brief description of what it was like in 1806
will be of interest. " It occupied about five
acres of ground, and its northern wall was
situated about fifty or sixty fathoms from the
sea beach. The walls were strongly built of
upright posts, with a wickerwork of reeds
between, supported from the inside by timbers
from six to nine inches in diameter, situated a
foot and a half distant from each other ; to
these the reed work was firmly lashed by tough
cinnet, made from the husk of the coconut.
The fencing was nine feet in height, but each
post rose a foot or so higher. There were four
large entrances, as well as several smaller ones,
secured on the inside by horizontal sliding
beams of the tough wood of the coconut tree.
Over each door, as well as at other places, were
erected platforms even with the top of the

fencing, supported chiefly on the inside, but projecting forward to the extent of two or three feet. These platforms were about nine feet square, and situated fifteen yards distant from each other ; they were used for the garrison to stand on, to shoot arrows, or throw down large stones, and, more particularly, to prevent a storming party from setting fire to the walls of the fortress. In front and on each side these platforms were themselves defended by a reed work six feet high, with an opening in front and others on either hand for the greater convenience of throwing spears, stones, etc. The lower fencing had also openings for a similar purpose. On the outside was a ditch nearly twelve feet deep, and as much broad ; this, at a little distance, was encompassed by another fencing similar to the first, with platforms, etc., on the outside of which there was a second ditch. The earth dug out of these ditches formed a bank on each side, serving to deepen them. In conclusion, the shape of the whole fortress was round, and both inner and outer fencings were profusely ornamented with white *pule* shells."

.   .   .   .

Immediately upon the arrival of Finau with his fleet in Nukualofa Harbour the expedition disembarked.   Eight of the fifteen Englishmen, with young Mariner, were armed with muskets taken from the captured privateer, and these at once opened a fire of musketry upon the enemy, who had sallied out of the fortress to oppose the landing.   So effectively were the eight muskets handled that Finau soon succeeded in landing his troops.   The first volley killed three and wounded several of the enemy, and a second threw them into such dismay that in five minutes only forty of the bravest remained to contest the landing, the rest retreating into the fortress.   In the meantime the seven other Englishmen had dismounted the *Port-au-Prince's* carronades from their carriages on the canoes and slung them to stout poles, and, conveyed by a number of natives, the guns were carried across the shallow water on the reef to the shore.   The rest of Finau's troops being then disembarked (4,000 in all says Mariner), the Englishmen again mounted the carronades, and a regular fire was begun at short range upon the fortress.

"Seated in an English chair taken from the

cabin of the *Port-au-Prince*, Finau took his station upon a portion of the reef not covered by the water, and watched the cannonade with intense interest. Much as he desired to lead his men to the assault, his chiefs would not allow him to expose his person by going any nearer. The fire of the carronades was kept up for an hour ; in the meantime, as it did not appear to do all the mischief to the exterior of the fortress (owing to the yielding nature of the materials) that the King expected, he sent for Mr. Mariner, and expressed his disappointment. The young sailor said that no doubt there was mischief enough done on the inside of the fort, wherever there were resisting bodies, such as canoes, the posts and beams of houses, etc., and that it was already very evident that the besieged Nukualofa people had no reason to think lightly of the effect of the artillery, seeing that they had already greatly slackened their exertions, not half the number of arrows being now discharged from the fort ; and, in his opinion, there were many slain lying within its walls.

"Finau was not satisfied, however, with his white artillerymen, but resolved to make an

assault, and set fire to the place, for which
purpose a number of torches, made from the
split spathes of the coconut palm, were prepared and lighted. An attack was then made
upon the first line of fencing and entrenchments,
which were, however, so weakly defended that
they were soon captured, and one of the door-
posts having been shot away, an easy entrance
was obtained to the inner fencing. This, in
many places, was not defended, and towards
these spots the storming party rushed with
lighted torches, whilst the enemy were kept
engaged elsewhere. The conflagration spread
rapidly on every side ; and as the besieged
endeavoured to make their escape, their brains
were knocked out by a second column of Finau's
troops, stationed at the back of the fortress for
that purpose. During all this time the Englishmen kept up a regular fire with unshotted guns,
merely to intimidate the enemy. At last a
general assault was made, and the conquerors,
club in hand, entered the place from several
quarters, and slew without mercy all they met
—men, women, and children. The scene was
truly horrible. The war-whoop shouted by
the combatants, the heartrending screams of the

women and children, the groans of the wounded, the number of the dead, and the fierceness of the conflagration formed a picture almost too distracting and awful for the mind steadily to contemplate. Some, with a kind of sullen and stupid resignation, offered no resistance, but waited for the hand of fate to despatch them, no matter in what mode ; others, that were already lying on the ground wounded, were struck with spears and beaten about with clubs by boys who followed the expedition to be trained to the horrors of war, and who delighted in the opportunity of gratifying their ferocious and cruel disposition. Every house within the great fortress that was not on fire was plundered of its contents, and thus, in a few hours, the fort of Nukualofa, which had obstinately and bravely resisted every attack for eleven years or more, was completely destroyed.

" As soon as Finau came within the fort, and saw several large canoes which had been carried there by the garrison, shattered to pieces by the round shot, and discovered a number of legs and arms lying around, and three hundred and fifty bodies stretched upon the ground, he expressed his wonder and astonishment at the

dreadful effect of the guns. Addressing his
men, he thanked them for their bravery, and
Mr. Mariner and his companions in particular
for the great assistance rendered by them.

"Some few of the enemy, who had escaped
the general slaughter, were taken prisoners.
They gave a curious description of the effect of
the guns. They declared that when a cannon
ball entered a house it did not proceed straight
forward, but went all round the place, as if
seeking for men to kill ; it then passed out of
the house and entered another, still in search of
food for its vengeance, and so on to a third, etc.
Sometimes one would strike the great corner
post of a house and bring it all down together.
The garrison chiefs, seeing all this dreadful
mischief going forward, rendered still more
tremendous by their own imaginations, sat in
consultation upon one of the large canoes just
mentioned, and came to a determination to rush
out upon the white men and take possession of
the guns. This was scarcely resolved upon
when a shot struck the canoe on which they
were seated and shattered it in pieces. This so
damped their courage that they ran for security
to one of the inner houses of the garrison, only

to see their men deserting them on all sides, and fleeing in terror from the dreadful round shot."

One incident in connection with this affair is also related by Mr. Mariner, who says that one of Finau's Fijian bodyguard, who had no doubt been present at the cutting-off of the *Port-au-Prince*, had taken from on board an earthenware fish-strainer, "such as is laid in the bottom of dishes when fish is brought to table. With this implement he had made himself a sort of breast-plate, and donned it at the assault upon the fort ; but unluckily it happened that an arrow pierced him directly through the hole which is commonly in the middle of such strainers." The wound laid him up eight months, and he never after-wards (in Mr. Mariner's time) was able to hold himself perfectly erect.

The last time that the fortress of Nukualofa came into prominence again in connection with white men was when Captain Crocker, of H.M.S. *Favourite*, was killed near there in 1842, when leading his men to succour the youthful King George, whose kingdom was in a state of rebellion. The result was most disastrous, for not only was he repulsed, but the rebels captured two guns as well.

" *Leviathan.*"

WHALING in the Southern Ocean and among the placid waters of the Pacific Isles may now be counted among the lost arts ; and yet well within the memory of many living men it flourished and was the main attraction that brought many a ship to the Southern Seas. And the brave and skilful whalemen of those days reaped such golden harvests from their dangerous toil, that those people who know not of these things would scarce credit the true stories that are told of fortunes made by whalers in the glorious days of the "thirties," "forties," and " fifties."

Yet, nowadays, whales are nearly if not quite as plentiful as they were then ; but where are the whalers ? And where is the once flourishing industry, and why is it now non-existent save

for two or three poorly equipped and manned
whaleships from Tasmania, and the shore whaling
parties of Norfolk Island? A glance at such
records as exist of whaling in the early days
makes still more remarkable its decadence at
this end of the century.

Even as far back as Dampier (1699) whales
were known to be plentiful in the two Pacifics.
" The sea is plentifully stocked with the largest
whales that I ever saw," wrote the scholarly
buccaneer. But this knowledge was not turned
to account until 1791 ; and this is how it came
about.

An enterprising London shipowner, named
Enderby, fitted out the *Amelia*, and sent her
round Cape Horn to endeavour to discover the
sperm whaling grounds. She left England in
September, 1788, and returned in March, 1790,
with 139 tuns of sperm oil on board. The news
spread, and a year later half a dozen American
whalers were cruising along the coasts of Chili
and Peru, and there was a great increase in the
quantity of oil imported into Great Britain.
Captain Phillip, who commanded the ships of
the "First Fleet," sent to Australia in 1788,
had also reported having seen a " vast number

of very large whales " on the passage out, and in July, 1790, while writing home on the vast potentialities for wealth that existed for those who would enter upon the business of whaling, mentioned that only a few days previously " a large spermacetty whale " had made its appearance in Sydney Harbour, capsized a boat and drowned a midshipman and two marines.

In the month of October, 1791, a convict transport named the *Britannia*, and owned by Mr. Enderby, arrived in Sydney Cove with her cargo of misery. She formed one of the " Third Fleet," and was commanded by a Mr. Thomas Melville. In the " Historical Records of New South Wales" there is a despatch from Governor Phillip which embodies a letter from Mr. Melville to his enterprising owners, the Messrs. Enderby, from which we learn that the *Britannia*, after doubling the south-west cape of Van Dieman's Land, "saw a huge sperm whale off Maria Island," but saw no more till within fifteen leagues of Port Jackson, when there came great numbers about the ship. " We sailed through different shoals of them from twelve o'clock in the day until after sunset. They were all round the horizon as far as we could see from the masthead. . . . I saw a very great prospect of establishing a fishery upon this coast." That was over a hundred and six years ago, and at the present time, during certain months of the year when the whales are travelling northward to the Bampton Shoals and the islands of the East Indian Archipelago, the same sight may be seen from any headland on the Australian mainland ; for to-day the whales are as plentiful and as fearless of human foes as they were then. But, alas ! the ships and the men are gone.

Melville, the master of the *Britannia*, was, however, a shrewd fellow, and as soon as he had got clear of his cargo of 150 convicts, he went to Governor Phillip and asked him to expedite his ship's departure so that he might cruise for whales even with the poor equipment he could secure in Sydney. He and his officers had tried to keep both their discovery and intentions in regard thereto a secret, but several of his ship's company were not so reticent, and when the *Britannia's* bluff old bows splashed into the sweeping billows of the Pacific four other ships followed almost in her wake. These were the *Matilda, William and Ann, Salamander,*

and the *Policy*, belonging to Hurry Brothers,
of London.   Reporting varying degrees of
success, the vessels (except the *Mary Ann*)
returned to Port Jackson in November and
December.   From that time the gallant Phillip,
in his despatches of 1791 and 1792, makes
frequent mention of the experiment, which he
justly considered was not a fair trial of the
Australian seas, although one reason he gave
for want of success was not correct.   He com-
plained that " the ships had not stayed out
long enough."   The real reasons, however,
were that while spermaceti and " right " whales
were plentiful enough, the whalers had not yet
learned their business, neither were they ac-
quainted with the creatures' migratory habits
in the Southern Seas, nor could they distinguish
between the profitable spermaceti, "right," and
"humpback " whale, and the dangerous and un-
assailable fin-back.   The *Britannia*, for instance,
ten days after her departure had seen, according
to her master, 15,000 whales, the greater
number of them off Port Jackson.   Now quite
two-thirds of this enormous number were the
swift and dangerous fin-back, a creature that,
while producing a certain amount of oil and

a small quantity of whalebone, is never attacked by boats, for it will tow a boat for thirty miles before it can be killed. While the shipmasters were agreed as to the vast number of whales, they considered that the bad weather and strong currents were obstacles too great to be overcome. However, they made other attempts along the Australian coast and then returned to England.

These same vessels, with many others, now became regular traders to New South Wales, bringing out convicts under charge of a military guard, and returning to England sometimes *via* China, would make a cruise to the " Fishery " before leaving the coast. Strange indeed were the adventures that befell the crews of some of these ships as they sailed northward through the islet-studded waters of the north-west Pacific, and no history of the sea would ever be complete that failed to tell these old and now almost forgotten tales of the mutinies, attacks by pirates, cuttings off by South Sea Islanders, and wrecks and disasters that are interwoven with the story of the British merchant marine in the Pacific from 1788 to 1850. Some of these wandering ships, unsuccessful in whaling, turned to sealing on

the coast of New Zealand. In the latter years
of the last century, however, the whalemen of
the time were gaining experience and a better
knowledge of the habits, feeding-grounds and
breeding resorts of both the right and sperm
whale, as well as of the great " schools" of
humpbacks and the less valuable flying finbacks, which made their appearance with such
undeviating regularity on the Australian coast
at certain seasons of the year; and slowly but
surely the business of whaling was becoming as
firmly established in the new colonies as it was
on the North American seaboard. Turnbull,
who made a voyage to New South Wales in
1798, and a voyage round the world in 1800–
1804, speaks of the growth of the industry
between the dates of his visits to New South
Wales. There were, he says, but four whalers
on the coast of New Holland in 1798, but at
the time of his second voyage there were fourteen, whose cargoes, on the average, "are not
less than from 150 to 160 tuns of oil, the
value of which at the present current price
amounts to between £180,000 and £190,000
annually.

Very early the Americans began to go south,

and in the old Sydney shipping records of the first years of this century there are many such entries as these : Arrived—*Favourite*, March 10, 1806, from Boston, America ; having refreshed, she sailed for the fishery. *Comanche*, from Juan Fernandez, with 300 barrels ; called to refresh, etc. etc.

The *Sydney Gazette* came into existence in March, 1803, and it was then and for several years the only newspaper in this part of the world. From its columns we learn that on February 14th " arrived the *Greenwich* whaler, Mr. Alexander Law, master, with 1,700 barrels of spermaceti oil, procured mostly off the north-east coast of New Zealand. The whalers she left cruising off that coast, and which may be expected here to refit about the beginning of June, are the *Venus*, *Albion*, and *Alexander*." The *Venus* duly arrived with 1,400 barrels of oil, and reported how her master had nearly lost his life when acting as harpooner, by the coil of the line getting entangled in his leg and dragging him overboard, but one of the boat's crew cut the line just in time to save his captain. From this date onward for a long time almost the only news

of importance in the convict colony is whaling
news, and that concerning the ships arriving
regularly from England bringing convicts or
stores.   These latter in most cases proceed to
the whaling grounds.   The ships as they come
in bring little scraps of news of the momentous
events happening in Europe at those times, and
the entries in the *Gazette* show us that the
whaling-men of those days had another element
of excitement and adventure in the lives they
led than that of encountering the whale.

For instance, in April, 1804, arrived the
barque *Scorpion*, Captain Dagg.   She " sailed
from England with a Letter of Marque the
24th of last June ; she has mounted fourteen
carriage guns, now in her hold, and carries
thirty-two men."

Most of the whalers at this time came out
armed with Letters of Marque, and more than
one vessel belonging to the Dutch settlements
was made prize to English whalers.   There
was the case of the *Policy*, which ship, in 1804,
was attacked by a Batavian vessel, called the
Swift.   Captain Foster, of the *Policy*, turned
the tables on the Dutchman, fought him for
some hours, took him prisoner, and brought

the *Swift*, which had once been a crack French privateer, his prize to Sydney. The story of the fight has been told in the old Sydney records, and it is not the only one of the kind which took place in these seas. Can it be that such episodes still linger in the traditions of the descendants of the Dutch settlers, and that the rankling of old wounds prompted the remarkable treatment of one Captain Carpenter, in the *Costa Rica Packet*—the one ewe whaling barque of Sydney—four or five years ago?

The whalers of those times had much to do with the discovery and exploration of the coasts of Australia and New Zealand, and deserters and men marooned from the whaleships began to settle on the islands of the Pacific—long before the missionaries were ever heard of.

Van Dieman's Land, or Tasmania, as it is now called, was already beginning to assume an importance in connection with the fishery. A *Gazette* of December, 1806, reports that Captain Rhodes of the *Alexander* whaler had arrived from the Derwent and Adventure Bay. He had about 100 tuns of oil, and a number of black swans from the River Huon. Rhodes told the interviewer of the day that he was of opinion that from the middle of May to the beginning of January the fishery in about the river Derwent would be very productive, a single vessel might procure 300 tuns ; after that season the weaning of the calves takes place, and the fish go north.

In a valuable work, " The Early History of New Zealand," published by Brett, of Auckland, New Zealand, the author of that portion of it from earliest times to 1840, Mr. R. A. A. Sherrin, gives some interesting particulars of the development of the whale fisheries. From this book we learn that, in 1808, whaling on the New Zealand coast was in a flourishing state, and that the *Grand Sachem* (Whipping, master) was about the first of a large fleet of American whalers which now began to frequent these waters. In the following year the *Speke*, Captain Hington, arrived in Sydney Cove with 150 tuns of black, and 20 tuns of sperm oil, this being the first recorded instance of the capture of the black whale.

The whaling grounds in the South Pacific are chiefly known as the " On Shore Ground," taking in the whole extent of ocean along the coast of Chili and Peru from Juan Fernandez

to the Gallapagos Islands ; the " Off Shore
Ground " the space between lat. 5 deg. S., and
long. 90 deg. and 120 deg. W. ; and the "Middle
Ground "—that between Australia and New
Zealand, and there are other grounds—the east
and west coasts of New Zealand, and across the
South Pacific between 21 and 27 deg. of S. lat.
The right whale fisheries are in the higher
latitudes in both hemispheres, which are the
feeding grounds, but as the winter approaches,
the cows resort to the bays to bring forth their
young, where they remain until the spring
months, when they again meet the bulls. Polack,
writing of sperm whales, says, " These fish are
gregarious, and migratory in their movements,
seldom frequenting the same latitude in an
ensuing season, and whalemen who have procured a cargo in one season, have often been
minus of oil by adhering to the same place in
the following year. No experienced South
Seaman will calculate for a certainty where he
will fill his ship. Those that have acted according to predetermination have returned to the
port they sailed from with scarce sufficient to
pay expenses."

In the " twenties " whaling had considerably

increased on the New Zealand coast, and in March, 1821, thirteen whalers had on board between them 6,960 barrels of oil. The increase of whaling soon led to complications with the Maoris, and quarrels ending in massacres begin to figure in the records. There is a tradition among the Maoris that the first Maori war arose owing to a dispute between two tribes, over some whales which were cast ashore on the coast.

In July, 1827, the Australian Whale Fishery Company was floated, and a year or two later, shore whaling began in Cook's Strait, and was soon followed in many of the bays on the New Zealand coast; while on the New South Wales coast, at Twofold Bay, a whaling station was a few years later established, and some little shore whaling is still carried on there to this day.

Benjamin Boyd, a Scotchman of good family, came out to Sydney in 1840 to take charge of some banking business, and in addition to many other speculations he went in largely for whaling, making Twofold Bay the rendezvous for his whaleships, and establishing a settlement known as Boyd Town. He was the first, or among the first, to employ South Sea Islanders, although,

of course, before this time whalers had often
a Kanaka or two among their crew. Boyd had
come out from England in a yacht called the
Wanderer, and in this vessel, owing to financial
disputes, he left the colony for California. On
his way he touched at one of the Solomon
Group, and it is supposed was murdered, as
reliable information has never been obtained as
to his fate.

In March, 1830, there were a dozen vessels
in the Bay of Islands, with 14,500 barrels of oil
on board, which, reckoning eight barrels to the
tun, gives a total of over 1,800 tuns, which was
at that time valued from £60 to £70 a tun,
the total value being then estimated at from
£111,000 to £130,000; and at the same time
a Sydney newspaper says :—" Three years ago
New South Wales had but three vessels engaged
in the sperm whale fishery, altogether about 450
tons, and the New Zealand trade was unknown.
She has now 4,000 tons of shipping engaged in
the sperm whale trade alone—and more than
9,000 tons of shipping have been entered out-
wards from this port for New Zealand, and
only since January 1st last to July 31st."
Early in 1831 the *Elizabeth* came into Sydney

Harbour with 361 tuns of sperm oil, the produce of an eighteen months' cruise, which was worth £22,000, the most valuable cargo of oil that had up to that time been brought into the port. A return of the exports from New South Wales for the years 1830 to 1840 inclusive, gives an idea of what a trade this was. In the return the exports are classified as sperm whale and black whale oil, whalebone, and sealskins ; without inflicting these figures on the reader it may be said that in 1830 the value of these exports was about £60,000, and each year the amount steadily increased, until in 1840 it reached £224,144.

Bay whaling in New Zealand had also increased to a considerable extent, the Maoris taking an active part with the Europeans in its development; but in New South Wales, with the exception of the settlement in Boyd Town, in later years the industry never established itself. Hobart Town has always been a regular port for whalers, and the industry survived longer there than at any place in the Southern Seas, although bay whaling had died out there by 1847. Norfolk Island was a regular calling place for the ships, and Lord Howe Island, with its population of sixty adults to-day, was originally settled by seamen from whale-ships, many of whom are still living—one, an American married to a Gilbert Island woman, has been upon the island for fifty years. At Norfolk Island boat whaling is now carried on by the descendants of the mutineers of the *Bounty*, most of whom, it will be remembered, emigrated there from Pitcairn.

So much for the beginning and heyday of whaling in the Southern Ocean ; the decline of the industry, and the causes of this, and the possibility of its revival, are worth discussing, but would perhaps be dull reading except to those of commercial mind. But the history of both the American and English whaling fleets are full of romance and daring adventure. Many and many a ship that sailed from the old New England ports and from Tasmania and New Zealand met with terrible experiences. Sometimes, as was the case of the *Globe* of Nantucket, they were cut off by the savage natives of the South Sea Islands, who, under the leadership of ruffianly beachcombers or escaped convicts, murdered every soul on board. Others there were who were actually

attacked by infuriated whales themselves and
sent to the bottom—like the *Essex* of London.
[Only a few years ago the writer saw in Sydney
Harbour the barquentine *Handa Isle*, which on
the passage from New Zealand had been so
attacked. She was a fine vessel of three
hundred tons, and was sailing over a smooth
sea with a light breeze when two large sperm
whales were sighted. They were both travelling
fast, and suddenly altering their course, made
direct for the ship. Then one sounded, but
the other continued his furious way and
deliberately charged the barquentine. He
struck her with terrific force just abaft the
mainmast and below the water line. Fortunately the barquentine was laden with a
cargo of timber, otherwise she would have
foundered instantly. The blow was fatal to
the cetacean, for in a few minutes the water
around the ship was seen to be crimsoned
with blood, and presently the mighty creature
rose to the surface again, beat the ensanguined
water feebly with his monstrous tail and then
slowly sank.] Some of these onslaughts upon
ships were doubtless involuntary ; as where a
whale, attracted by the sight of a ship, had

proceeded to examine her, misjudged his distance, and came into collision with disastrous effect to both. But there are many instances where the whale has deliberately charged a ship, either out of pure "devilment," or when maddened with the agony of the wound inflicted by a harpoon.

Many years ago, a small school or "pod" of sperm whales was sighted off Strong's Island, in the Caroline Archipelago, by the ship *St. George* of New Bedford, and the Hawaiian brig *Kamehameha IV*. Both ships lowered their boats at once, and in a very short time Captain Wicks, of the Hawaiian brig, got fast to a large bull who was cruising by himself about half a mile away from the rest of the "pod." As is not uncommon among sperm and humpbacked whales, the rest of the school, almost the instant their companion was struck, showed their consciousness of what had occurred, and at once crowded closely together in the greatest alarm, lying motionless on the surface of the water as if listening, and sweeping their huge flukes slowly to and fro as a cat sweeps its tail when watching an expected spring from one of its own kind. So terrified were they with the

knowledge that some unknown and invisible
danger beset them, that they permitted the
loose boats—five in number—to pull right on
top of them.   Four of the boats at once got
fast without difficulty, leaving three or four of
the whales still huddled together in the greatest
fear and agitation.   Just as the fifth boat got
within striking distance of the largest of the
remaining fish, he suddenly sounded, and was
immediately followed by the others.   Some
minutes passed before Martin, the officer of
the fifth boat,. could tell which way they had
gone, when the *St. George* signalled, " Gone to
windward ! " and presently Martin saw them
running side by side with the whale which had
been struck by Captain Wicks.   Martin at once
started off to intercept them, and when within
a few hundred yards he saw that the stricken
whale was surrounded by four others, who
stuck so closely beside him. that Captain
Wicks could not get up alongside his prize
to give him the first deadly lance thrust
without great danger.   At last, however, this
was attempted, but the whale was not badly
hurt, and the four other fish at once sounded
as they smelt the creature's blood.   But, sud-

denly, to Martin's horror, the huge head of an enormous bull shot up from the ocean, directly beneath the captain's boat, the mighty jaws opened and closed and crushed her like an eggshell! Fortunately Wicks and crew sprang overboard the moment they caught sight of the creature's fearful head, and none were killed, although two were seriously injured. Martin at once picked them up. Meanwhile the cause of the disaster darted away after his three companions and the wounded fish which was lying on the surface spouting blood (much to Martin's satisfaction, for he feared that the infuriated creature would destroy his boat as well as that of Captain Wicks). The condition of the wounded men justified Martin in making back to the ship, and he at once gave orders to that effect, seeing that another boat from the brig was hastening to kill the wounded whale. Hastily putting Captain Wicks and his men on board his ship, Martin again started out to meet 'the four loose whales which were now coming swiftly down towards the ships. The big bull which had destroyed Wick's boat was leading, the others following him closely. Suddenly, however, he caught sight of Martin's

boat, swerved from his course and let his companions go on without him. Then he lay
upon the water motionless as if awaiting the
boat to attack and disdaining to escape.

But just as the boat was within striking
distance and Martin had called to the harpooner, " Stand up ! " the whale sounded, only
to reappear in a few minutes within twenty
feet of the boat, rushing at it with open jaws
and evidently bent upon destroying it and its
occupants. So sudden was the onslaught that
Martin only saved himself and his crew from
destruction by slewing the boat's head round as
the monster's jaws snapped together ; but as
leviathan swept by he gave the boat an " under
clip " with his flukes and tossed her high up in
the air, to fall back on the water a hopelessly
stove-in and shattered wreck. And then to the
terror of the crew as they clung to the broken
timbers, the whale returned, and the men had to
separate and swim away, and watched him seize
the boat in his jaws and literally bite it to pieces,
tossing the fragments away from him far and
wide. Then after a minute's pause, he turned
over and began swimming on his back, opening
and shutting his jaws and trying to discover his

foes. For five minutes or so he swam thus in widening circles, and then as if satisfied he could not find those he sought, he turned over on his belly again and made off. Almost immediately after Martin and his men were rescued by another boat.

But the whale had not finished his work of destruction, and as if goaded to fury by the loss of his companions and the escape of his human foes, he suddenly appeared twenty minutes later close to the Hawaiian brig. He was holding his head high up out of the water and swimming at a furious speed straight toward the ship. The wind had almost died away, and the brig had scarcely more than steerage way on her, but the cooper, who was in charge, put the helm hard down, and the whale struck her a slanting blow, just for'ard of the forechains. Every one on board was thrown down by the force of the concussion, and the ship began to make water fast. Scarcely had the crew manned the pumps when a cry was raised, "He's coming back!"

Looking over the side, he was seen fifty feet below the surface, and swimming round and round the ship with incredible speed, and

evidently not injured by his impact. In a few
moments he rose to the surface about a cable
length away, and then, for the second time,
came at the ship, swimming well up out of the
water, and apparently meaning to strike her
fairly amidships on the port side. This time,
however, he failed, for the third mate's boat,
which had had to cut adrift from a whale to
which it had fastened, was between him and the
ship, and the officer in charge, as the whale
swept by, fired a bomb into him, which killed
him almost at once. Only for this he would
certainly have crashed into the brig and sunk
her.

Another well-known instance was that of the
ship *Essex*, Captain George Pollard, which was
cruising in the South Seas in May, 1820, and
which is related as follows : " The boats had
been lowered in pursuit of a large school of
sperm whales, and the ship was attending them
to windward. The captain and second mate
had both got fast to whales in the midst of the
school, and the first mate had returned on board
to equip a spare boat in lieu of his own, which
had been stove in and rendered unserviceable.
While the crew were thus occupied, the look-

out at the masthead reported that a large whale was coming rapidly down upon the ship, and the mate hastened his task in the hope that he might be in time to attack it.

"The whale, which was a bull of enormous size, and probably the guardian of the school, in the meantime approached the ship so closely that, although the helm was put up to avoid the contact, he struck her a severe blow, which broke off a portion of her keel. The enraged animal was then observed to retire to some distance, and again rush upon the ship with extreme velocity. His enormous head struck the starboard bow, beating in a corresponding portion of her planks, and the people on board had barely time to take to their boat before the ship filled and fell over on her side. She did not sink, however, for some hours, and the crew in the boats continued near the wreck until they had obtained a small supply of provisions, when they shaped a course for land ; but here, it is to be regretted, they made a fatal error. At the time of the accident they were cruising on the equator, in the longitude of about 118° west, with the Marquesas and Society Islands on their lee, and might have sailed in their boats to either of these groups in a comparatively short time. Under an erroneous impression, however, that all those lands were inhabited by an inhospitable race of people, they preferred pulling to windward for the coast of Peru, and in the attempt were exposed to and suffered dreadful privations.

Those few who survived their complicated disasters first made the land at Elizabeth or Henderson's Island, a small, uninhabited spot in the South Pacific, and which, until then, had never been visited by Europeans. After a short stay here part of the survivors again put to sea in search of inhabited land, and ultimately reached the coast of South America; and an English whaler at Valparaiso was sent to rescue those left on the island, but found but two alive."

Unless the writer is mistaken the *Handa Isle* was not the first vessel bound from New Zealand to Sydney that was struck by a whale; for about twenty-eight years ago a small barque named the *King Oscar* met with an exactly similar experience, and could not be kept afloat. Then, in the year 1835, "the ship *Pusie Hall* encountered a fighting whale, which, after injuring and driving off four boats, pursued them to the ship, and withstood for some time the lances hurled at it by the crew from the bows of the vessel before it could be induced to retire."

An Island King.

TILL within eight years ago the comman-
ders of the various ships of war on the
Australian station used to be familiar with the
name of an island king who, in his small way,
gave them considerable trouble. He was, perhaps, the most famed of all the chiefs of that
vast area of scattered islands and islets in the
North and South Pacific vaguely described as the
" South Sea Islands," and, indeed, his courage in
war, cunning in diplomacy, and general all round
" cuteness," were only equalled by the famous
old Samoan fighting chief, Mataafa, and the late
Maafu, the once dreaded Tongan rival of King
Cacobau of Fiji. The name of this personage
was Tem Benoke, but he was generally known
as Apinoka ; and his dominions were the great
chain of coral islands which enclose the noble

lagoon of Apamama, the largest but one of the recently annexed Gilbert Islands. The population of these islands, comprising the atoll of Apamama, is now something over a thousand, and they do not show any signs of diminution —probably owing to their disinclination to accept the introduction of European civilisation and a sudden change of habits and mode of life generally.

For nearly fifteen years Apinoka ruled his people with a rod of iron. All the revenue derived by his subjects from the sale of their produce, such as copra and other island commodities, was paid into the Royal treasury, and from there it found its way into the pockets of trading captains, who sold the aspiring King modern breech-loading rifles of the latest pattern. As time went on he began to harry the people of the neighbouring islands of the Gilberts, and soon threatened to be the one dominant ruler of the whole group. And then the missionaries—native teachers working under the supervision of the Boston Board of Missions —began to get alarmed. For missionaries in general Apinoka ever expressed the most withering contempt, and word went out that any of his people who accepted Christianity would have his life cut short. And, as Apinoka was ever a man of his word, the people of Apamama obeyed.

A hundred yards from the white beach that faced the inner and eastward side of the lagoon he built his state house, a cool, airy building of semi-European design and construction, and here he sat day after day, surrounded by his Danites—grim, black-haired, and truculent—dictating his commands to his American secretary and another white, his interpreter and chief cook, Johnny Rosier. All round the spacious front room were boxes and cases of all sorts and descriptions of island trade and merchandise—tins of biscuits, kegs of beef, cases of gin, sardines, salmon, and piles of old-fashioned muskets and modern rifles.

But although he kept a white secretary and interpreter, the King did not like white men. He had once bought a schooner, and although he engaged a white captain and mate, and paid them liberally, he treated them otherwise with scarcely disguised contempt. They were necessary to him— that was all ; and at any moment his dark, heavy face might put on a dangerous

look towards these venturesome whites. Then
it was time to clear ; to leave the island absolutely, for he would tolerate no white man
living on shore except those actually in his
service and in his favour.

.   .   .   .   .

Back from the house were the copra sheds
and other buildings used for storing the King's
produce ; and all day long his slaves toiled about
him, cutting up the coconuts and drying them
on mats in the fierce, hot sun. Patiently and
in silence they worked, for they knew that those
small, keen eyes, under that heavy, sullen brow,
might fall upon them if they rested or talked.
And then might the King give a sign, and one
of the guards would come with a weighty stick,
and the sound of savage blows upon naked
backs be heard.

Out upon the broad, shady verandah sat the
Royal harem—women captured mostly from
Apian and Tarawa and Maiana. Before them
was spread a profusion of food—native and
European—and as they ate and talked in low
whispers each one sought to rival the others in
her caressing attentions to a strong, handsome
boy of ten years of age who ate among them.

For this was the King's adopted son, and the apple of his eye. Children of his own he had none, and this child of his brother's was the one object of affection in his savage mind.

Presently the boy—a spoiled and petted tyrant (he is now King)—strikes one of the women a rude blow on the cheek, and desires her to haste and bring his bath towel—he would bathe in one of the King's fresh water fish ponds. One by one the women of the harem rise to their feet, and with bent shoulders and downcast eyes pass the huge figure of their dreaded owner. For the boy must not be let to go alone for his bath. Perchance an old coconut might drop from a tree and fall near him if he wandered alone, and that would mean a sudden and bloody death for them. So one by one they file away through the groves of palm trees, the boy, pipe in mouth and towel on arm, leading the van.

No one speaks to them as they pass through the village, and only the women may gaze at them ; the men, especially if they be young and stalwart, turn away their faces in silence till they pass. For perhaps a sly glance might pass, or an idle word be spoken, and then some day an evil tongue might whisper that Teran, the King's toddy-cutter, had said to a comrade that Nebong, the tenth wife of the King, was good to look upon. Then would men come to Teran's house in the night, and call to him to rise and come with them, and then as he walked with them along the darkened path a knife would gleam, or a shot ring out, and Teran be heard of no more. Neither would his name again be spoken, unless in a whisper, among those of his own kith and kin.

.   .   .   .   .

So in this way went on the days on surf-girt Apamama, and Apinoka the King grew fat and waxed strong, and the terror of his name, and cold, merciless nature reached from Arorai in the south, to Butaritari in the north. But, by-and-by, there came about a rumour that all this steady buying of rifles and revolvers, and ammunition, meant ill for the people of the islands to the south, and many of the white traders, who hated the grim old despot, joined hands with their hereditary foes, the native teachers, and made common cause together for his downfall, which soon came about, for one day a British gunboat steamed into the lagoon.

A message was sent to the King to come aboard, and with it a threat to make no delay, else matters would go ill with him. So in his white helmet-hat and black suit the King came off, rowed to the gunboat in his own whale-boat. With a glum, stolid look upon his face, and savage rage in his heart, he was helped up over the ship's side, and escorted to the cabin, and in five minutes more he knew his power of conquest over other islands was gone for ever.

But before a word was spoken on either side, the King walked over to the chair that was at the head of the table, and, with a snort of mingled bodily relief and defiance to the naval officer, managed to squeeze his huge body into a sitting posture.

"Get up out of that chair, you confounded savage," said the captain, sharply. "What do you mean by sitting down there? Squat on your hams, like the thundering savage you are," and he pointed to the cabin floor. "You have no missionary or trading captain to deal with now."

Slowly he rose, fixing his eyes in wondering rage upon the unmoved face of the

officer. Then he squatted cross-legged on the
floor.

" Every gun, every pistol, and every cartridge on the island must be brought on board
this ship," were the startling words he next
heard. No use was it to try to coax or
wheedle this captain or tell him lies ; and then,
while the King remained on board in sullen
silence, men were sent to collect the arms.
From that day forth the *mana* of Apinoka
weakened, and then, although the bulk of
his people stood loyally by him in his days
of trouble and paid their tribute as of yore,
there were many who gave voice openly to
their hatred, and to their joy at his downfall.

A year or so passed, and the King, sitting
in his grand house, and looking across the
waters towards the islands of Kuria and
Aranuka, whose people his forefathers had slain
in bloody massacre, grew daily more sullen
and savage as he thought of his vanished
glories, swept from him by the hated white
man. A small lump that had formed on one
of his huge legs began to pain and irritate
him, and so the native doctors were called,
and he commanded them to cut it open. He

was no common man, to be dosed with medicines like a sick woman. "Cut," he said.

They cut, and in twenty-four hours Apinoka was at his last gasp. Calling his head men and his harem around him, he commended his boy nephew to their care. "Let him be king in my place," he said, and then, not deigning to say farewell to his numerous wives, who wept around him, he took a draw at his pipe, and went to join the other monarchs in the spirit world.

A Spurious Utopia.

TWO years or so ago, with a wail of sorrow and indignation from its people, Norfolk Island came to the end of its existence as the Utopia of the balmy Southern Ocean, and, sentimentally speaking, was wiped off the map. For henceforth its future will be the care of the practical-minded Government of New South Wales ·and instead of the delighted visitor to this dreamful isle being welcomed on the shiny strand by youths and maidens garlanded with flowers and chanting a melody of welcome, as is generally supposed to be the island custom, he will be met by hotel-runners and other prosaic evidences of a practical civilisation. But the Norfolk Islanders bitterly resented the change forced upon them. For forty or more years they had been patted on the back by the world in general and romancists in particular as " an English-speaking and deeply religious community, brave, fair to look on, generous, and virtuous." Also they were said to be endued with a large selection of minor qualities, to enumerate which would take up a goodly-sized catalogue. Books innumerable have been written by enthusiastic globe-trotters (mostly ladies) on the beautiful Arcadian existence that has ever been the lot of the islanders since the British Government took compassion on the cramped condition of the rapidly multiplying descendants of the famous *Bounty* mutineers on lovely little Pitcairn Island and removed two hundred of them from there to Norfolk Island in 1856. Long before that time, however, the Pitcairn Islanders were held up—and very deservedly so— to public admiration as an ideal community, and their future career in their new and beautiful home was watched and read about with the deepest interest. In a very short time, however, primarily inspired by their intense affection for Pitcairn, some of the emigrants became dissatisfied and rebelled against the conditions of life in the new and more spacious paradise ; and no less than sixty

of them determined to return to their beloved
isle, that "lonely mid-Pacific rock, hung with
an arras of green creeping plants, passion
flowers, and trumpet vines ; and breasting back
the foaming surf of a mighty ocean." But
the paternal British Government did not take
kindly to the idea, and instead of sixty only
seventeen people were allowed to return to
Pitcairn. These were families of the Youngs,
descendants of Edward Young, the comrade
of the ill-fated Fletcher Christian. The
seceders consisted of two men, their wives,
ten girls, and three boys ; and to this day
they and their descendants, augmented by an
occasional dissatisfied McCoy, or Adams, or
Quintal from Norfolk Island, dwell in peace
and comfort on their old island home three
thousand miles away from the rest of the
Bounty stock. Sometimes a wandering trading
vessel brings news of them, but, even in
Australia, Pitcairn is all but forgotten. Not
so, however, with the Norfolk Island people—
who, by the way, are all Youngs and McCoys,
and Quintals and Adamses, and Nobbses and
Buffets (the first four being family names of
their *Bounty* progenitors). They always were

.and are now, very much in evidence. Scarcely a year has gone by since 1856 but some traveller has written of their open-hearted generosity, their simple piety, and their daring courage as whalers. Their life was depicted as ideal, their loyalty to the Queen as something touching, and they almost said cross words to one another when contending for the privilege of entertaining a visitor. The use of liquor was unknown, and its name abhorred ; a wicked word, even when some stalwart whaleman darted his iron at a whale and made a shocking bad miss, was unheard of, and there was no record of even the words " cat " or " nasty thing " being applied to each other by the dark-eyed island beauties, even under the strongest provocation. It was a home of unearthly bliss and strict rectitude of conduct.

Time went on and the population steadily increased to its present numbers—about 600. The Melanesian Mission established a training school for its young native ministers, and the fame of the people was noised abroad, and their happy lot was the theme of many a pen and the inspiration of many an author. Then came whispers of discontent. Strangers, called "interlopers," had settled on the island, and were not satisfied with the patriarchal and family system of government. They wrote letters to the outside world—and talked. And they also openly asserted that the morals of the *Bounty* descendants were not as good as they were supposed to be. Then came dissensions among the community generally, and fierce quarrels among the *Bounty* families as to certain rights and privileges ; and out of all this came certain statements which gave a shock to the ordinary common Christian of the outside world. No one believed any evil, however, of the Norfolk Islanders for a long time ; but at last rumour became so strong, and the unchristian "interlopers" made such distinct charges against them, that the Government of New South Wales intimated to the people that an investigation into the administration and condition of the island was desirable. The Norfolk Islanders rose up as one man and protested in a loud voice against such an indignity. For forty years they had basked in a world-wide reputation for unblemished goodness, and why should they be "investigated"? And least of all would they submit to be investigated by the Government of New South Wales. They, as a community, were Great Britain's one ewe-lamb of a spotless life in the South Seas, and no one but the Queen herself had the right of having them " investigated." New South Wales, they admitted, had some sort of nominal authority over them ; but they were not going to tolerate anything like an official investigation by a colonial government.

The Government of New South Wales, however, was obdurate, and two years ago, despite the angry protestations of the majority of the islanders, a Commission was sent down from Sydney empowered to make a searching inquiry into the administration of the island laws, and to ascertain whether or not it would be advisable to administer Norfolk Island according to the laws of New South Wales.

For a fortnight the Commission held nearly daily sittings, and examined a great number of witnesses, and then at the conclusion of its labours the President, Mr. Oliver, called a public meeting of the male inhabitants, and addressed them very eloquently. He pointed out to them that they had no power to disregard the laws made for them by the former Governors of New South Wales and substitute laws of their own, and that their continual maladministration even of their own so-called laws had at last brought trouble upon them. He did not want, he said, to say hard things ; " but," he continued, " you have been sadly misrepresented by people who have visited you for a short time — due no doubt to your hospitality to them." And then he told them something more unpleasant still. ". . . The rottenness of their condition was very evident . . . the island is in a most deplorable condition . . . crime is rampant and unchecked . . . the morals of the younger people are as low as they can possibly be." The island, he pointed out, was supposed by the world generally to be a home of smiling plenty, and that the moral and social condition of its people had no parallel, whereas the very reverse was the case. Their lazy habits had been a curse to the island, and the condition of the land, as compared with what it was when the place was turned over to them in 1856, was deplorable indeed ; it was simply becoming the home of the poison bush and the wild tobacco plant. They imagined—

and had imagined for forty years—that their proper policy was to exclude strangers, as they had done at Pitcairn Island. But that was a mistake. They were not capable of taking care of themselves, "and for their own welfare it was eminently desirable that colonists should be admitted to the island." And then the President of the Commission pointed out the beneficial results that had attended the establishment of the Melanesian Training Mission on the island, and concluded his address by a kindly appeal to their common sense to remember that the glaring maladministration of justice and the utter disregard by the island authorities of instructions sent to them by the Governor of New South Wales had alone brought about the interference of the Colonial authorities.

At the conclusion of the President's address, which was received in sullen and astonished silence, the medical officer of the island, a man universally respected, proposed " That it be represented to the Governor of New South Wales that it is *not* desirable for Norfolk Island to be annexed to New South Wales." To this the Commissioner made a brief but emphatic answer. He declined to allow such a resolution—in the face of the results of the Commission's investigations—being put.

And those who know the kind-hearted, hospitable people, and the splendid agricultural capabilities of their island for earning its place as one of the gems of the Pacific, will be sincerely glad of such a radical change. Its resources will be developed and its social conditions vastly improved under the new *régime*, which by simply pulling away the veil of sentiment that has so long enwrapped the Norfolk Islanders in a spurious reputation of possessing all the virtues, will transform its inhabitants from being useless into good citizens of the Empire.

Love and Marriage in Polynesia.

THE rapid advance of civilisation and the spread of Christianity for the last fifty years among the Malayo-Polynesian races of the South Pacific have had, naturally enough, much to do with either the partial abandonment or the total extinction of many of their customs. In some cases this, and the substitution of European for native habits, is to be regretted, such as, for instance, the quick and inconsiderate adoption of European clothing by a people whose daily habits of life and constitution rendered them peculiarly unfitted for such a sudden and violent change. Between 1823 and 1830, when the natives of Rarotonga and other islands of the Cook Group, following their

chiefs' example, abandoned their heathen practices
for Christianity, the most terrible mortality resulted from the ill-advised action of the missionaries inducing their converts to clothe
themselves *en masse* as a practical proof of their
spiritual change, and an outward and visible
sign of grace. Precisely the same result has
attended the introduction of Christianity in the
Marshall and Caroline Groups by the American
missionaries. Nowadays, however, in this respect
a more liberal conception of the laws of nature
and health is possessed by missionaries in general
all over the world than was the case in the earlier
years of the present century. Then—and were
it not for the pathetic side of the question, one
might be inclined to laugh at such inconceivable
folly and ignorance being displayed by educated
men—it was thought essential for a convert,
who, perhaps, had for fifty years worn nothing
more than a waist-girdle of pandanus leaf or of
thin calico, to be garmented in a suit of heavy
black cloth or woollen material, and adopt as
well the habits and manners of civilised life.
That many thousands of people died from
pulmonary complaints engendered by this
sudden change, the natives themselves assert ;

and, indeed, only a few years ago the people of one of the North-Western Pacific Islands almost entirely succumbed to pulmonary disease caused by their wearing heavy clothing during the rainy season. Previously, when they wore nothing more than a simple waist cloth or girdle of grass, such diseases were absolutely unknown, but their desire to resemble white men as much as possible, and the earnest supplications of the resident Hawaiian teacher who implored them to dress as he did, in cloth, proved fatal to these simple-minded people.

Among other customs that have undergone a rapid change, or have been altogether discontinued, is that of marriage according to the old rites and ceremonies, with its many interesting and often pleasing details. In all those islands —except Samoa, perhaps—that have been the scene of missionary labours, the ceremony of marriage is now performed by either a white missionary or native teacher, and is a very prosaic affair, divested as it is of all the old attendant feastings and merrymakings. But among the Micronesian race inhabiting many of the scattered islands of the Western Caroline Group, the old native customs have as yet scarcelv undergone any great change, and the ceremonies attending the marriage of any chief of note are as prolonged and imposing as are the dances and other festivities, which last for some weeks.

Ellis, who made a careful study of the manners and customs of the Malayo-Polynesians inhabiting the Society, Austral, and Hawaiian Islands, gives some very interesting particulars of the marriage customs during the early days of missionary enterprise in the South Seas. At the present day these are unknown, and, indeed, the younger generation of natives are almost as ignorant of the customs and practices of their forefathers as a Yorkshire labourer is of those of the people of Tierra del Fuego.

In Samoa, although the commoner people are married in the European fashion in a church, the higher chiefs still cling with pardonable tenacity to many of the old practices observed in former times ; and, indeed, so strong a hold has the observance of such ceremonies upon the Samoan mind, that while the poorer classes are content to be married according to the rites of the Christian religion, they eagerly enter into all the preparations for the celebration of a chief's marriage with the ancient rites, which generally

precede the subsequent ceremony performed by a
white missionary or native teacher.

In Tahiti, the celebration of marriage, says
Ellis, took place at an early age, " with females
at twelve or thirteen, and with males at two or
three years older (and, indeed, this is still the
practice). Betrothment was the frequent method
(as it is in Fiji at the present time) by which
marriage contracts were made among the chiefs
or higher ranks in society. The parties themselves were not often sufficiently advanced in
years to form any judgment of their own, yet,
on arriving at maturity, they rarely objected to
the engagements their friends had made."
Sometimes, however, previous attachments had
been formed, which resulted in the same tragedies
that occur from the same cause in civilised life. I
remember hearing of one such instance which occurred at Niué, the " Savage Island " of Captain
Cook, only a few years ago. A young native girl
had become much attached to a man who, with a
number of other islanders, had gone away under
a two years' engagement to work the guano
deposits on Howland Island, in the Equatorial
Pacific. When they returned she learned that
her lover was dead, and from that day her once

gay and merry demeanour left her. She fell into a deep melancholy, and confided to two of her girl friends that her parents, now that her lover was dead, insisted upon her marrying another suitor, whom she regarded with indifference, if not dislike. One of her confidants, a girl of about eighteen, had been for many years afflicted with a painful disease in the bones of her left foot, and suggested to her friend that they should both end their sorrows by suicide. The third girl, who was the youngest of all, earnestly sought to dissuade them from such a deed, but, finding her pleadings were unavailing, said she would not remain alive to lament their loss. They seemed to have made their preparations for death with the utmost calmness and fortitude, and, dressing themselves in their best, they leaped over the cliffs, and ended their lives together.[1]

In Tahiti and the other Society Islands the period of courtship, in Ellis's time, " was seldom protracted among any class of the people; yet all the incident and romantic adventure that

[1] PUBLISHER'S NOTE.—This incident is related in detail in " Pacific Tales," under the title of " For We were Friends Always," by the same author.

was to be expected in a community in which a high degree of sentimentality prevailed, occasionally came to pass, and the unsuccessful suitor was sometimes even led to the commission of suicide, under the influence of revenge and despair. Unaccustomed to disguise either their motives or their wishes, they generally spoke and acted without hesitation ; hence, whatever barriers might oppose the union of the parties, whether it was the reluctance of either of the individuals themselves, or of their respective families, the means used for their removal were adopted with much less ceremony than is usually observed in civilised society." As an instance of this, he relates the following authentic story : A young chief of Murea (or Eimeo), an island a few miles from Tahiti, became attached to the niece of one of the principal *raatiris*, or landowners, on the island of Huahine. He was one of the body-guard of Taaroarii, the king's son, and although only twenty years of age, was already distinguished for his courage in warfare and his gigantic stature and perfect proportions, while his pleasing countenance and manners and engaging disposition generally, rendered him a favourite with both whites and natives. The girl's family admitted his visits and favoured his designs, but the object of his choice declined every proposal he made. No means to gain her consent were left untried, but all proved unavailing. He discontinued his ordinary avocations, left the establishment of the young prince who had selected him for his friend more than his servant, and repaired to the habitation of the girl he was so anxious to obtain. Here he appeared subject to the deepest melancholy, and, leaving the other members of the family to follow their regular pursuits, from morning to night, day after day, he attended his mistress, performing humiliating offices with apparent satisfaction, and constantly following in her train whenever she appeared abroad. His friends interested themselves in his behalf, and the disappointment of which he was subject became for a time the topic of general conversation in the settlement among natives and whites alike. At length the young lady was induced to accept his offer. They were publicly married, and there being nothing of the New Woman about this Polynesian beauty, lived very happily together. Their married life, however, was but of short duration, for

his wife, for whom he appeared to cherish
the most ardent affection, died a few months
afterwards.

Later on an instance of another kind occurred. A party of five or six persons arrived
in a canoe from Tahiti on a visit to some
friends in Huahine, one of the Leeward
Islands. Their original destination was Bora-
bora, but they remained several weeks at Huahine, the guests of a chieftainess named Teraimano. During their stay, a young woman of
great beauty, " one of the belles of the island,
and who belonged to the household of their
hostess, became exceedingly fond of the society
of one of the young men, and it was soon intimated to him by some of her girl friends that
she wished to become his companion for life.
The intimation, however, was disregarded by
the young man, who expressed his intention of
prosecuting his voyage—like the lover who
kisses and rides away. The girl made no
secret of her distress, and her beauty of face
and figure suffered such a remarkable change in
so short a time that her friends became deeply
concerned ; but she yet showed her preference
for the object of her affection by scarcely

leaving his side. But finding that the young fellow, who was barely past eighteen years of age, was unmoved by her attentions, she not only became exceedingly unhappy, but declared that if she continued to receive the same in-difference and neglect, she would either strangle or drown herself. Like the Niuéan girl before-mentioned, she had friends who sought to dis-suade her from her purpose; but as she declared her determination was unaltered, they used their endeavours with the stranger, who afterwards returned the attentions he had received, and the couple were married at Huahine. His com-panions pursued their voyage, and afterwards returned to Tahiti, while the newly-married couple continued to reside with the chieftainess Teraimano. Their happiness, however, was of short duration ; not that death dissolved their union, but that attachment which had been so ardent in the bosom of the young woman before marriage was superseded by a dislike equally as powerful, and she subsequently treated her youthful husband with insult and contempt, and finally left him."

In 1882, when I resided on Mădurŏ, one of the Marshall Islands, a young man, Jelik, a brother of the chief of the district, conceived an ardent affection for a young woman who was employed as a servant by a German trader named Wolff.   She was of foreign blood, being a native of Arrecifos, or Providence Island (North-West Pacific), from whence she had been brought by the trader during her childhood. Possessed of ample means, the young man sought to show his affection in the most extravagant manner by making the girl presents of all sorts of articles, both European and native.   Among his gifts to her was a handworked sewing machine—just then coming into use among the natives of the Marshall Group— and a keg of salt meat.   Both of these were bought from a white trader at a high price— about five times their English value—and were subsequently bought back by him (the trader) from the girl for a few dollars.   Her object in selling them was, she said, to make her lover a present.   The money she at once expended in the purchase of tobacco and a small clasp-knife ; and her lover, instead of being, as would be imagined, angry at her conduct, expressed the greatest delight at receiving such a proof of her regard. A few months later it was my happy privilege to be present at the marriage and assist Jelik in receiving two or three other white men who were invited to be present. At the conclusion of the marriage ceremony—which was performed according to ancient custom, for the missionaries had not then succeeded in making any converts on Mădurŏ — the bridegroom announced his intention of putting away his two other wives, whom he had hitherto treated with respect and affection. This, however, the young lady from Providence Island strenuously besought him not to do; and although barely sixteen years of age, she made an eloquent appeal to her husband before the assembled guests, and declared that she would at once return to the protection of her former master's wife rather than consent to such an injustice. Her extreme youth, she said, would not allow her to supersede in such a sudden and cruel manner two women who had never done her an injury; she would rather dwell in accord with them under their joint husband's roof and be taught by them in her wifely duties than subject them to an outrage and do violence to her own feelings. Her earnest appeal to her husband softened him, and he consented to retain his two former wives, explaining to his newly-wedded one and the white traders, that while he had no cause of complaint against the original sharers of his married life, they were women of no rank or position, and would not themselves feel aggrieved had he persisted in his intention. His conduct was in no degree singular in this respect, and the three wives got on very happily together afterwards. Like the people of the South Pacific Islands, however, the woman to whom a Micronesian chief or person of distinction is first united in marriage is generally considered as the head of the establishment, and although he may subsequently marry one or two more wives of higher rank than the first, she would hold a superior position to the new-comers.

Niué: the "Savage Island" of Captain Cook.

THREE hundred miles eastward from the Friendly Islands, and rising abruptly from the blue waters of the Pacific, is the lonely and verdure-clad Niué, the "Savage Island" of Captain Cook, and the abode of one of the most interesting and conservative peoples in Polynesia.

If you make the island anywhere on its northerly or easterly coast, you will not like its appearance. Before you lies what seems to be a rounded mass of floating green, the base hidden from view by a misty haze that may be either fog or smoke. But as the ship gets well into the land, and point after point opens out, you see that the cloudy mist is neither smoke nor fog, but the spray of the wild surf beating unceasingly against the long, monotonous line of grim and savage-looking cliffs that rear their dreaded fronts from Makefu to Fatiau. All day long, be the sea as smooth as glass oceanwards, or be the trade wind gone to sleep, the narrow ledge of black and jagged coral reef that here and there juts out at the foot of the forbidding wall of grey is smothered in the boil and tumble of the restless breakers; and where there is no shelving reef to first arrest and break their fury, the huge sweeping seas race madly inward, and with the roar of heavy artillery fling themselves in quick and endless succession against the face of the perpendicular cliffs, to pour back in sweeping clouds of snowy foam. Sometimes, if the south-east trade is blowing lustily, the roar and crash of the surf seems to shake and vibrate the coral wall to its foundations, and the thick and matted scrub that lines the summits of the cliffs to their very verge is drenched and flattened by the sheeted spray, and the swaying fronds of the coconut-palms growing further back from the shore are wetted and soaked by the lighter spume.

There is no barrier reef to Savage Island, and consequently no harbours. Anchorages there are—one at Avatele and one at Alofi, the two largest towns—but even these are only available during good weather, and when the trades are steady. Many a good ship has met her fate on the cruel shore of Niué, and among them was the second *John Williams*, missionary ship of the London Missionary Society; she was wrecked there in 1867. And long, long before the first wandering white man ever landed on the island, unknown ships had run ashore there, and never a soul was left to tell the tale; for in those days even had the sea spared the lives of the castaways, the spears and clubs of the ferocious natives would have made quick work of them. And even nowadays, when every native on the island is a decided Christian, and goes to church twice a day on week-days and four times on the Sabbath, they candidly admit that they do not like white people, and only tolerate their presence for benefits derivable from intercourse with them.

Though the island is but forty miles in circumference, there are over five thousand

natives living in the eleven villages that are
situated at pretty wide intervals on Niué, and
although, since the introduction and adoption
of European clothing, pulmonary and other
dreaded diseases have become prevalent, the
population shows no signs of decreasing ; in
fact, it has shown a slight increase since 1872.
Before describing the people, however, it should
be mentioned that Niué is one of the few of those
peculiarly-formed islands known as " upheaved
coral," and although the interior is either a
series of impenetrable guava scrub interspersed
with belts of heavy timber, coconut groves,
and masses of jagged coral rock covered with
a matted growth of vine and creeper, the
decomposed coral soil is of wonderful fertility,
and, given one condition—an industrious people
—this solitary and little-known island would
be one of the richest in all the South Pacific.

Five years ago I first saw Niué and afterwards
spent six months there. A very stormy passage
from Tonga had thoroughly sickened me of
the hideously dirty and uncomfortable trading
steamer in which the voyage was made, and it
was delightful to hear the rattle of the cables
through the hawse-pipe that told us we had

reached our destination—the village of Avatele, which was to be my home. We were anchored so close to the shore that I could hear from my cabin the shouts and cries of the natives as they gathered together on the rocks awaiting the boats to land, and I hurried to dress myself and my little daughter so as to get ashore in the first boat.

The scene from the deck of the ship was a pretty one. Between rocky headlands there lay a tiny little beach—the only one on Niué— from which a rough path led to the village, an irregular cluster of brown thatched houses standing among lofty coconut-palms ; and further back on a level greensward, white buildings of coral lime contrasted prettily with the wealth of the dark green foliage of orange and breadfruit trees that grew around them. Beyond, nothing was to be seen but an endless array of the greyish-red trunks of the graceful coco-palms that encompassed the village on all sides but the sea front.

By the time we had taken our seats in the boat, the whole of the village had gathered together on the rocks—men, women, and children. I had only just time to notice that all the women were dressed in long gowns of the brightest colours—red, green, blue, scarlet, and indeed of every other hue imaginable ; and that their long, coarse black hair hung loosely down upon their backs like horses' tails, when the boat touched the landing place, and the noise, which had been bad enough before, now became simply indescribable ; and then, before I could recover my dazed senses, we were fairly rushed by hundreds of women and girls, who fought and struggled with each other for the privilege of shaking hands with the " new " *papalagi* (white man) and his child, who had come to live among them.

Up the rocky path we were borne towards the house of one of the resident white traders, whose guests we were to be till my own house was put in readiness. Presently we reached his gate, and here there was a mad rush to get inside. My little daughter, who was close behind me, carried by a pleasant-faced woman named Hakala, began to get terrified at the deafening noise and excitement. A short, muscular-looking young native with a light-brown skin and dandified black moustache, pushed through the women, knocked them aside with scant ceremony, and made room for me to get inside the fence and on to the verandah of the house. Then a pretty, pale-faced, little white lady—the trader's wife—came out and welcomed me warmly to Niué.

Outside the fence the swarm of gaily-clad women and children shrieked and yelled at "Nikolasi Tane" and "Nikolasi Fafine" (literally "Nicholas and his wife"), not to take the new white man and his child inside just yet ; they wanted to *kitia* (look at) them a little longer. And then they tried to force their way in, despite the angry verbal remonstrances of "Nikolasi Fafine," and the good-natured but hearty punches and kicks administered them by Soseni, the native teacher's muscular son. At last our hostess carried us off in triumph to her comfortable sitting-room (her husband was busy landing some trade goods) where I met the rest of her family, the younger members of which, although in manners and appearance exactly like other English chil-dren, only spoke and were spoken to in the native language. This, my hostess explained to me, was an inevitable consequence of long residence in the islands, and although the children were actually educated in English, it was impossible to get them to talk in anything but Niuéan — even to their parents — and then, naturally enough, the parents themselves answered them in the same tongue.

In a few weeks or so I was fairly settled down to the routine of life in Savage Island, and began to take an interest in the people. Candidly, they are not nice people—not by any means. In appearance the Niuéans are a strongly-built, muscular race, darker in colour than the Samoans, and without many of the good qualities that distinguish the latter race. In Samoa you cannot walk about anywhere in the villages without the natives calling out and asking you to come inside out of the heat of the sun (*Sau i fale ma le la!*) and drink a coconut. In Niué you may ride or walk all round the island on a blisteringly hot day and meet, perhaps, fifty natives of either sex carrying bundles of drinking coconuts, but they will walk stolidly past, unless you happen to have some tobacco to give them. Then you will get a drink and, if the piece of tobacco you tender is not big enough, the man or woman you give it to will not hesitate to tell you that you are *lamakai*—shockingly mean. In Samoa, at night time, fires are lit and mirth and merriment prevail, and the sound of singing and dancing may be heard in every village after evening service. In Niué there is none of this; there is no dancing—that is strictly forbidden, save a decorous, semi-religious performance that takes place when a new church is opened, or on the occasion of some religious function; and as for singing, nothing but hymns are tolerated, hymns shouted at the very greatest tension of naturally harsh and guttural voices in a tongue that is a curious combination of Maori and Hawaiian. As soon as darkness falls upon the islands the natives retire to their dwellings. No one, unless it is some enterprising fowl or pig stealer, or a stealthy lover hurrying to his trysting-place, will brave the darkness. Now and then you may hear the crunching of the broken coral pebbles in the roadway, and the tread of footsteps of the *leo leo*, or policemen on their beat to arrest any one who is abroad without good reason.

. . . . .

At daylight the people are up and about. Those who own plantations of yams, taro,

sugarcane, or bananas, set out to work before
the heat of the sun gets too great. But they
must be back in time for morning service.
Others—and these are in a majority—will loll
outside the trader's store door waiting for him
to open, and here they will lie and loaf about
half the day, buying nothing themselves, but
watching the people from other villages bring in
their baskets of copra, sea-island cotton, fungus,
bundles of arrowroot, vegetables, fruit, and
other native produce to sell to the white man.

One monotonous day succeeds another, only
to be broken by the cry of " Sail, Ho ! " Then
the village wakes up, and for the next two or
three days the wildest activity prevails. After
the ship has gone the white traders and their
wives visit each other in succession, and hear
or tell the latest news from Sydney or Auckland, for month after month has passed and no
ship has come. Perhaps there is a *fono* (the *tapu*
of other islands) on the coconut trees, and no
copra can be made for six months ; and the
cotton is not yet ripe for picking. Then the
trader knows what *ennui* means. He has read
all the books on the station, and life becomes a
weariness. There are no white, sandy beaches on

Niué, laved by placid lagoons, where one can walk for miles, as in other islands ; no pigeons to shoot in the dense, scrubby interior, and only one thing that can be done, and that is saddle his horse and ride round the island. For there are horses on Niué, and fairly good ones too, although they are terribly tender-footed, owing to the rough nature of the country. There is but one road on the island, which, starting at Avatele, winds its sinuous and erratic course among the groves of coco-nuts that fringe the rocky coast right round the island.

I shall always remember my first ride there. The station horse was an old New Zealand hurdle racer, which had been taken over by an Island trading firm for a bad debt owing them by some unfortunate. He was sent to Niué, and because of his alleged habit of bolting had acquired an evil reputation, and I was earnestly cautioned not to ride him. He could never be trusted, I was told, and many terrible calamities would happen if I tried it. I would be killed before I had gone a mile. I was not. He walked quietly out of the station gate, and undisturbed by the cries of the natives, the yelping of the teacher's dogs, and the grunting and squealing of the scores of pigs that lay basking in the sunshine of the narrow road, he trotted along over the crunching pebbles till the "street" of Avatele was left behind. Then, once his tender feet felt the soft red soil beyond, he cantered gaily along till the first obstruction was reached, a high fence of coconut logs, erected across the road to prevent the village pigs wandering into the bush. The sound of the horse's feet brought a rush of people to the narrow gate. They fought and swore violently at each other as to who should open the gate for the white man. How good of them. Alas! no. They charge for politeness in Niué. A stick of vile, strong-smelling tobacco is the fee for opening any gate. If you have not got it with you, you will have to give them a written I O U for it. Most likely the bearer of the order will give it to a friend who has a bruised finger or a cut foot, who will swear you out that *he* was the man who opened the gate, and that in so doing a log fell on his hand, or the horse trod on his foot and cut it, and demand another stick of tobacco for compensation.

A mile or so from Avatele the road turns off at the village of Tamakautoga, and ascends the plateau, and here for a mile or two is a lovely bit of verdant tropical beauty—an avenue of shady palms, interspersed with orange and lime trees. Then comes a flat, sandy plain covered with patches of guava scrub and native plantations of sugarcane. Sometimes, where the road passes through a guava thicket, the ripe guavas fall about the horse as he pushes the branches aside with a toss of his head. Six miles from Avatele and you catch a glimpse of blue sea now and then through the dense foliage, and come to the edge of the plateau before the road descends to Alofi ; and then two hundred feet below you can see the open coast and the steep coral cliffs again, and hear the roar and thunder of the ever-beating surf.

No one wants to go further than Alofi the first day, for at Alofi is the home of a man who, with his amiable and hospitable wife, has, during his five-and-twenty long years of unceasing toil on Savage Island, endeared himself not only to every trader on the island but to every wandering white man, be he captain or fo'c'sle hand, who has ever stood under his roof-tree.

And there is one thing to be said of the Niué
native ; and that is, that, with all his faults, he
would give his life for the white missionary
who is not only his teacher and adviser in things
spiritual, but his doctor, his protector, and his
friend.

.   .   .   .   .

"White men lead such a lazy existence in
South Seas, do they not ? " is a question often
asked, and usually answered in the affirmative.
But there are exceptions to every rule, and the
white trader, and his white or native wife on
Savage Island, do not lead the dreamy, careless,
and lazily happy sort of life which Herman
Melville has written in those charming books—
"Typee" and "Omoo."   Not that he is kept
continuously busy all the year round, for it
sometimes happens that the native rulers place
a *fono* upon the coconut trees, and during the
period that the *fono* (the *tapu* of other islands)
is in force, which may be from one to six
months, the business of copra-making ceases,
and although there is much other island produce
to be bought, such as arrowroot, fungus, and
cotton, these form but a comparatively minor
adjunct to the mainstay of the island trade,

which is copra. At the time of my arrival there were five traders on the island, who, while absolutely yearning for each other's society during the slack season, were mortal enemies, from a business point of view, during the copra season. Business competition was very keen, and the natives took advantage of it to the fullest extent by demanding such a high price for their copra and cotton that the white men had to combine and agree as to a maximum price. After a hard battle with the natives, the latter yielded and peace was restored.

At our village, Avatele, there were two traders. There were also two at Alofi, and one (the *doyen* of the vocation on Niué) at a distant village called Mutulau. As there are eight other towns, besides those mentioned, which have no resident traders, the people of these eight places have to carry their produce in some cases nine or ten miles. Thus, to Avatele would be brought copra, arrowroot, fungus, and cotton from the large towns of Hakupu and Liku, distant six and eleven miles, as well as from the nearer small towns of Fatiau and Tamakautoga, while Alofi and Mutulau were the markets for Uhomotu, Tamalagau, Tamaha-

tokula, and Makefu. Sometimes parties of two
hundred or three hundred natives would arrive
at Avatele from, say, Hakupu, each man and
woman carrying two baskets of copra slung on
a pole and weighing, say, 80 lb. or 100 lb. They
would generally start on their journey long
before daylight and reach Avatele before the
sun's rays grew too powerful. Likely enough,
they would find that another large party of
people had come in from Tamakautoga on a
similar mission, and had taken possession of all
the available ground surrounding the traders'
stores.

Now, the rankest jealousy between the various
towns obtains on Niué, and, consequently, in a
few minutes, wrangling and fighting would
begin, the women taking an active part in the
proceedings. For half an hour or so the noise
would be deafening; then the bell for morning
service rang and quiet reigned till its close.
By this time the white men had finished breakfast and were ready to open their stores and
commence the day's trading. And a day's
trading on Savage Island during the copra
season is enough to try the temper of a saint.
Let me try and describe it.

The two rival trading stations at Avatele almost adjoin each other. Each trader has his own particular adherents, and long before he is ready to throw open his store these have brought their baskets of copra and dumped them down against his door. Naturally enough the Avatele natives try to be first in the field, and block the people from outside villages from getting too near the door for the first rush. Perhaps the trader has just opened a case of something lovely in the way of prints —a green and yellow check upon a brilliant scarlet ground—and the Avatele women are determined that none of that print shall go to hated Hakupu. Each man is accompanied by his wife, and each wife is accompanied by as many of her female relatives as she can muster; also her children. If she has no children there are always plenty of volunteers, and every one —men, women, and children—mean to handle that piece of print, and prevent any outsider buying it. By-and-by the whole of the vacant ground, stretching from the traders' stores away up to the white-walled native church, is covered with hundreds upon hundreds of excited people, every one of whom has from two to a dozen

baskets of copra to sell, and is possessed of a
set resolve to get it weighed and sold before
any one else.

As the sun gets higher, the impatience of the
waiting, wrangling crowd increases. And still
more people come in by various paths leading
from the interior of the island. All these, too,
have heavy loads of the rank, oily-smelling
copra, packed in large baskets made of plaited
coconut leaves. Generally the load is slung to
a pole, the ends of which rest upon the naked
shoulders of the bearer. As they stagger down
the rocky path that leads from Fatiau and
Hakupu, they are greeted either with cries of
welcome or jeers from those who have arrived
before them. With a cry of relief their burden
is dropped, and then from the baskets and the
bearers' backs and shoulders arises a black swarm
of flies. Flies are one of the two curses of
Niué. The other is the curse of grass seed.
The latter only troubles the white people's
garmented legs ; the former make no distinction between white man or native. Leaving the
darkened and fly-protecting shade of your house
and going out into the dazzling sunshine you become black with flies in five minutes. They crawl

into your ears and settle in your eyes. Brush
them off and kill them, and for every hundred
you slay a thousand cheerfully buzz into their
place. You meet a native. He looks like a
perambulating figure composed of flies. As he
passes he gives himself a vigorous brush with
a branch he carries. You do the same. Two
black clouds arise and assimilate and then divide
forces. If the native is a bigger man than you,
he gets most.

.   .   .   .   .

At last the trader has finished his breakfast
and makes for the door of his *fale koloa* (store),
and a roar of approval comes from the natives,
and then ensues a wild stampede. First of all,
though, he looks out through a peep-hole and
calls out *fui tau lago* (" Brush off your flies ").
The women set to work and strike out vigorously
right and left. The men will most likely call
out to the trader to come and brush them away
himself (I think I have mentioned before that
your Savage Islander is not a Chesterfield).
The door opened, the trader steps into the
breach. A huge platform scale is wheeled out
in front of the counter, beside which he takes
his stand, note-book in hand. If he is a single

man he will weigh, say, some 10,000 lb. or 20,000 lb. of copra, and then leave off weighing and go behind his counter and pay for it before weighing any more. If he is married, his wife pays for each lot as it is weighed. Before he proceeds to weigh the first lot, however, he may call out—" Do any of you people here think I want to cheat with this *fua* (scale) ? "

Immediately some one will answer, " Yes."

" *Mitaki* (good). Then let some of you come up and try the scale." He does not get angry —not unless he is new to the cheerful candour of the Niué people.

A basket of copra is brought up and placed on the scale. The trader weighs it, and apparently takes no notice of some half a dozen natives who stand by with pencils and note-books. They are missionary pupils—*i.e.*, sucking ecclesiastics. He knows just as well as they do that that basket of copra has been weighed half a dozen times by as many native teachers, and its weight carefully noted. Every teacher has a steelyard, and every bag of cotton, or fungus, or basket of copra, that goes out of any village is weighed on that steel-yard before it is sold to a white man. Born cheats themselves, they trust no one.

Silence for a few seconds, and then the trader calls out, " *Siau ma tolu pouna* (103 lb.)."

A sigh of relief comes from the natives, and the six young ecclesiastical gentlemen with pencils murmur, " *E tonu* " (correct).

" Three pounds off for the basket," says the trader, as he hands the seller an I.O.U. for the amount due to him. A howl of rage, and then a chorus of such expressions as " Robber, *tagata kolea* (bad man), *pikopiko* (liar). 'Tis a shame to say it weighs 3 lb. It weighs but 1 lb."

"All right," says the trader placidly, "capsize it, and let us weigh the basket." The basket is a thick, heavy one of green coconut leaves, made purposely heavy, and weighs just 6 lb. "Here, give me back that bit of paper, and the trader scratches out the figures 100 and writes 97 instead.

Probably the seller will swear, if not at the trader, at himself. Then he gives place to some one else. At the end of three or four hours the white man calls out that he is tired and hungry, wants to eat something, and his dinner awaits him. If the natives are in a good temper, they grumble good-naturedly, and tell him not to bother about his dinner, they will send some

one to eat it for him. If they are cross, they
will tell him that gluttony is the curse of all
white men, and suggest that if he cannot find
time to attend to his business he had better
give up trading altogether.

So the day goes on, till darkness brings an
end to the noise and work, and the wearied
white man, with hands and face smothered in
greasy copra dust, goes back to his dwelling-
room for a prefunctory wash and to eat a
hurried meal. He has given out, say, five
hundred I.O.U.'s, ranging in value from 50c.
to 20 dol. each, and many of the holders of these
are people from distant villages who must be
paid that night, as they want to get back.
So, lighting some lamps, he opens the store
again. Already there is a swarm of people
waiting. The first I.O.U. is handed in by a
woman. He looks at it—400 lb. at 2c. per lb.
—8 dol.

"What do you want for this?" he asks.

"Six fathoms of *ie vala vala* (muslin)."

"Yes; go on," and he ticks off 1 dol. 50c.

"*Salu piko* (a poll comb)."

"Yes," and he marks down 25c.

"*Fagu Maskala* (a bottle of musk)."

" Right, a dollar.   Come, hurry up, don't go
to sleep ; what else ? "

" *Isa*, saucy white man.   How much is left ? "

" Five dollars and a quarter."

" Have you any *fulu fulu kula* (scarlet ostrich
feathers for head adornment) ? "

" Yes ; here you are, a dollar each.   How
many ? "

" Two."

" Three dollars and a quarter left.   Come, *fia
moe koe* (are you going to sleep) ? "

The woman laughs good-naturedly as she
caresses the curling scarlet feathers admiringly.
" Can I have some money ?   I want a dollar for
tau poa nè mè (the missionary offering)."

" Yes.   Here you are.   That leaves one dollar
and a quarter.   What else ? "

" Nothing now ; *fakamau* (place it to my
credit)."

" Right.   What is your name and place ? "

" Talamaheke, from Fatiau."

The trader enters her name in a book, tears
up the original I.O.U. for 8 dol., and gives her
1 dol. for the amount remaining to her credit.
And so on till he is too tired to do any more,
and shuts up his store till the following morning.

The Old and the New Style of South Sea Trader.

THE old style of trader has disappeared from the South Pacific, though north of the Equator he is still to be found. Thirty years ago he lived like a king, and called no man master, and, save the then slowly-growing missionary interference in his domains, had nothing to trouble him. A year or so ago in an English paper there appeared a long article on "the traders of the South Sea Islands," wherein the old style of men were described as being nothing short of a band of red-handed cut-throats and inhuman savages. Their days were spent in shooting indiscriminately all those who crossed them, and their nights in the most hideous debauchery and drunkenness. They were *always* "either escaped convicts, or felons who had evaded justice by fleeing beyond the bounds of civilisation," etc. To those who knew anything at all of island life as far back as twenty-five or thirty years ago, this description was of interest alone from one point of view—it displayed a remarkable ignorance of the subject. That there were in those days some few notorious scoundrels and villains of the deepest dye living in the South Seas was true enough, but they were few, and very far between.

Nevertheless, some very silly things have been written in the past, and will no doubt be written in the future, about many of those wandering and adventurous, yet honest men, who made the Pacific Islands their home from the days of Wallis, Cook, and Bligh. Certainly during the *régime* of the terrible Convict System of New South Wales some notorious desperadoes did escape to and live in the South Seas. They were generally of the worst type of convicts, men whose hands were against every man's, for every man's hand was against them ; but they were but few in number when compared with the legitimate traders who had established themselves on hundreds of islands

from Pitcairn in the east to the Ladrones in the
north-west.　　That many of these men were
deserters from the great fleet of whaleships which
from 1798 up to 1850 cruised through both
Pacifics is also true, but they were not the terrible
villains it was the fashion to describe them.
The greater number of them were simply
traders, and sold coconut oil, pearl-shells,
and provisions to the whaleships.　　Some of
them made fortunes ; others merely made a
living—but a living free from the hardships
and miseries of a life at sea ; some there were
who became thorough natives, and could not be
distinguished from the wild people among whom
they lived and died.

　　But let me give an idea of the life led by
both classes of the old-time island adventurers—
the genuine runaway sailors who had become
traders, and the escaped convicts from Van
Dieman's Land or New South Wales.　　Fifty
years ago, with these latter the cutting-off of
ships was a favourite pursuit, and the capture
of the *Globe*, of Nantucket, was a notorious
instance.　　The *Globe* lay at anchor in Milli
Lagoon in the Marshall Islands, when an
escaped Tasmanian convict and a Portuguese

named Antonio Gomal, who were living on
the island, planned with the natives to capture
the ship and massacre the crew. This was
successfully accomplished, but it is satisfactory
to know that Mr. Gomal at least did not enjoy
his victory, for the ship's cooper, seizing a
harpoon, sent it through the " Portingal's " body
and pinned him to the deck-house. Then, a
few years later, a band of thirteen convicts living
on Pleasant Island—an isolated spot in o deg.
25 min. S., 167 deg. 5 min. E.—and aided by
two hundred natives, cut-off a ship whose name
was never ascertained, and murdered every soul
on board. This ship was well armed, and her
crew of fifty men made a determined resistance.
She was then plundered and burnt, and in
the hilarious festivities that ensued on shore to
celebrate her capture, two of the white men,
Goad and D'Arcy, got drunk and shot a chief
who claimed more than was considered a fair
share of the booty. In an instant a general fight
ensued, five of the whites were slaughtered by
the chief's retinue, and the remaining eight,
with their native wives, were compelled to leave
the island. They were never heard of again,
although the boat in which they left was

picked up some months later at an island
a thousand miles to the westward ; in it
were the remains of three of the poor women.
So much for one class of the old island adven-
turer.

But if there were a few such ruffians scattered
about the then little-known groups of islands in
the North and South Pacific, the generality of
the old style of traders were a good stamp of
men ; and much of the success that has attended
the labours of the missionaries in many parts of
Micronesia and Polynesia is due to the influence
these unknown and forgotten men exercised
over the natives. The earlier missionaries
on many islands had much to contend with
in the bitter opposition of many of the white
traders, who resented their interference in the
domestic relations of the whites with the natives,
yet in many instances they were sensible of the
fact that only for the presence of these adventurous whites Christianity would not easily have
gained a footing in the Caroline and Marshall
Islands.

Still, the idea that all, or nearly all, of the old-
time traders were men who had broken the laws
of their country, and had fled from justice, and

were of desperate deeds and licentious habits, seems to be pretty generally accepted even to the present day, and such belief dies hard where romantic adventure is concerned. Many of them, it is true, had to carry their lives in their hands, and take life to save their own. One such man perhaps is still alive. Fifty years ago, when still a young man of twenty-six he was shipwrecked on one of the Caroline Islands, and up to 1876 had roved about from group to group with his numerous children and grandchildren. On one occasion, when living on Kusaie Island, he found that a plot existed among some of his followers—who were all related to him—to seize a small trading vessel that lay at anchor in one of the harbours of the island. The ringleader of the plot was one of his sons-in-law, a native of the Kingsmill Group. This man revealed his murderous designs to his wife, who, in her turn, communicated them to her white father. Arming himself, and accompanied by half-a-dozen of his half-caste sons, the old trader at once boarded the schooner, told her captain of the plan to seize the vessel and kill all hands, and said that he would, if the captain assisted him, mete out justice. Two boats were manned and armed, and went ashore, and the trader, calling his followers together in the village square, demanded the names of those who had planned the seizure of the schooner. Nine men were named, among them his son-in-law. No attempt was made at denial, and the unfortunate wretches, together with six Caroline Islanders who had joined in the plot, were marched to the beach and summarily shot. "If he (the leader) had been one of my own sons, instead of only my son-in-law, and had been proved guilty, I would have had him shot," he said, when speaking of the occurrence. And yet this stern old fellow was not only respected, but loved by his native followers, none of whom, not even those who were themselves concerned in the plot and were shot for it, would have thought of questioning his authority even in a matter of life and death.

Some of those bygone traders lived in great style. They generally were sufficiently astute to marry a woman of some social rank and position, by which means their own status and influence with the natives was sure to be increased. Notably this was the case in the older days in the Caroline and Marshall Islands, where some two or three white men equalled in power and position the highest chiefs in the land. On Ponapé, in 1820, one such man maintained a force of some hundreds of fighting men, all of whom were armed with muskets and cutlasses. He had originally been the master of a trading and whaling vessel, which also did a little quiet privateering, during 1815–20. Chased away from the East Indies by Dutch and English men-of-war, he was sailing eastward to try his luck against the Spaniards on the coast of South America when he lost the vessel at Ponapé in the Western Carolines. Out of his crew of thirty men, nearly twenty returned to China in a small schooner they built from the wreck for the purpose; while he and the remainder accepted the offer of the principal chief of the Jakoits district to stay on the island and assist him in his warlike expeditions. Like the survivors of the massacred crew of the English privateer *Port-au-Prince*, who assisted the warlike chief Finau to subjugate the whole of the Tongan Islands, these eleven adventurous seamen went to work with such zeal that in six months the three districts of Ponapé were under the sway of the chief who had secured their services.

As might have been expected, most of these men in due time met with violent deaths. Two or three of them, however, remained with their leader, and like him, although in a lesser degree, became rich and powerful, and lived on the fat of the land. What their names were is not known, but the memory of their doings has not yet died away in the Western Carolines. Their leader married the widow of the king of one of the subjugated districts of the island, and choosing the south end of Ponapé for his domain, lived in state. Wherever he went he was attended by his body-guard, and these he subjected to a rigid military discipline. He exacted a small tribute of tortoiseshell and coconut oil from the people of the district he had conquered; but also bought much of the same articles from them for export to China. He was eventually lost at sea in a small vessel he had himself constructed at Jakoits Harbour, and was mourned by a great number of light-coloured people of whom he was the progenitor.

Another old trading identity was Harry T——, who died a few years ago. He had

formerly served in the English navy, and, in
addition to imparting much useful knowledge
to the natives of Pleasant and Ocean Islands,
taught the particular tribe with whom he lived
the use of firearms, with the result that Pleasant
Island, long notorious for the continual bloodshed that was always occurring between the
seven clans or tribes that inhabit the island,
settled down to comparative peacefulness ; for
Harry threatened to exterminate the other six
clans and divide the place among his own
adherents unless they gave up warfare. A hale,
stalwart old fellow, by no means devoid of religious feeling, he was a type of man-o'-war's man
of the days of Nelson—always ready to fight, but
yet brimming over with kindly impulses towards
whites and natives alike. His dwelling-house
and store was the general rendezvous not only
for his own very numerous native and half-
caste following, but for the few other white men
trading on the island. He was looked upon
as a father to the community, and it was a
matter of pride with him, whenever a ship
called at the island, to invite the captain
ashore, and, after treating him royally, point
to a withered old woman of seventy, and say,

"That's my wife, sir. I married her nigh on forty years ago, an' she's been with me ever since. An' I ain't agoin' to put her away—as the custom here is, when a wife gets old." He always took great pleasure in showing visitors how his sons could box, and would often, old as he was, put on the gloves with one of his stalwart boys, "to keep him from gettin' rusty." Nine sons and seven daughters, most of whom were married, and had families, made the old man's dwelling a credit to him when they were all under the same roof; and the sincere respect and admiration they all evinced towards the patriarch might well have made him feel proud.

Another ex-man-o'-war's man, who lived on Upolu, in Samoa, had the distinction of possessing about the largest family and the hardest pair of knuckles of any white man in Polynesia. He was much esteemed by the natives for this latter fact, as well as for the open contempt with which he treated the mandates of the then British Consul to appear before him and answer charges of assaulting Germans and other foreigners when making one of his periodical visits to Apia. At last, however, a stray cruiser

from the South American station happened to
call at Samoa for provisions, and the irate
Consul took advantage of her presence. A boat
was sent up to the old man with a request for
him to come down and appear at the Consular
Court. Nothing alarmed, he cheerfully complied, and dressed in white " ducks," and
attended by half a dozen of his sons and a
numerous train of natives, he made his appearance in due time at the court. There were
present a good number of white residents, some
of whom were Englishmen, who had come to
watch the proceedings, and others, who were
foreigners, to bear witness against the fighting
proclivities of the old trader. Seated beside
the Consul were the captain and doctor of
the man-of-war ; and behind them the American and German Consuls, both of whom much
wished to see the old man punished.

Several charges of assault and battery were
then proceeded with, and three Germans and
two other foreigners related how the old fellow
had knocked them about " for nothing at all."

" Nothing at all ! " said the trader furiously,
and, turning an appealing look upon the spectators, he was about to give his version of the

affair, when the Consul stopped him and told him " to hold his tongue."

Had a thunderbolt fallen at his feet the old fellow could not have shown greater surprise. For a moment he gazed at the consular representative of Britain's power and might with a ludicrous expression of mingled amazement, anger, and contempt ; and then stepping up to the captain of the warship, he addressed him.

" Look here, sir. You don't know me, and you might think I'm lying. But I'm not. Ask any man here—except any of these miserable Dutchmen " (indicating the German Consul-General for the Pacific), " or a blatherskitin' Yankee like *him* " (nodding at the American official), " or a shuffling old henwife like *this* apology for an Englishman " (pointing his finger at the British Consul)—" ask any one here, I say, if this fussy ass would have dared to tell me to hold my tongue if the captain of a man-o'-war wasn't here ? " And he turned wrathfully to the amused assemblage to corroborate his remarks.

Trying hard to restrain a smile, the naval officer advised the old man not to interrupt, and to " treat the court with respect."

"Respect!" And the clean-shaved, wrinkled features of the ex-man-o'-war's man so darkened with rage and contempt that the naval officer's smile of amusement gave place to a look of concern. "Respect a man that will be a party to let a lot of blessed furriners insult me—me, an English sailor—in the streets of this town, and then, because they can't fight, lay a complaint agin' me in the Council's (Consul's) office. Look here, captain, I served with Admiral Cochrane in Chili, and I has a proper contempt for all furriners. I'm an old man, an' an old fool—I gets drunk whenever I comes to town, and these Dutchmen insults me by sittin' down by 'emselves an' a-lookin' at me—me, an Englishman—as if I was a naked kanaka. An' whenever they does that I gets up an' plugs 'em."

"But that won't do, B——" said the naval officer, severely. "You must not take too much to drink, or you cannot keep out of trouble."

"Well, sir, I can't forget I'm an Englishman. An' this here Council has no more pluck in him than a cat. He's fined me an' fined me over and over again for usin' what *he* says is insultin' language to a lot of furriners, who, if it wasn't for the likes o' me, would make every Englishman in these here islands ashamed of his country."

For a moment or two the captain of the man-of-war wore a troubled look upon his face. Then he held up a warning finger to the trader.

"Now just listen to me, sir, and let me warn you against assaulting your foreign fellow-residents. And I advise you to respect your Consul. If you do this I shall take good care that no injustice is done to you."

The old sailor's face brightened, and with a defiant look at the three Consuls he raised his hand and saluted.

"Right you are, sir. If you thinks I've done wrong in pluggin' half a dozen miserable furriners for insultin' me, I'm willin' to pay my fine like a man. But, sir, I believe that a dozen fellers like me could lick both watches on a German frigate."

Struggling hard to keep his countenance at the old trader's earnest manner, the naval officer discussed the matter with the three Consuls, and the ex-man-of-war's man was fined five dollars.

Then an American storekeeper, who had

been a much interested and amused spectator
of the proceedings, advanced to the Consul's
clerk, and, placing a sovereign upon the table,
turned to the old trader.

" Charley, old man, I'll pay your fine. This
has been most enjoyable."

But the old order of things in the islands of
the North and South Pacific is changing rapidly,
and ere another score of years have passed the
last one of the old style of traders will have
disappeared. The new style of trader is merely
a shopkeeper, pure and not simple, for he buys
and sells over a counter, and keeps books, and
carries an umbrella, and only for his surroundings might be taken for a respectable suburban
grocer in England. Of course, however, if you
sail away beyond the usual track of the regular
trading vessel even to-day you will come to
places where a scanty few of the old style of
men still exist in their isolation. But these are
men who have made money in the older times.
They have pushed out in disgust from the
civilised and crowded groups of Eastern Polynesia, where the voice of the tourist is now
raucous in the land, because the new conditions
of life became hateful to them, and the incessant

competition that assailed them in their business lessened their prestige with the natives more and more every day. And so, as the new men came in and opened " shops " in Fiji, and Tonga, and Samoa, and the Hervey Groups, and the sailing vessels were replaced by dirty, frowsy-looking steamers with loudly dressed supercargoes, who came ashore with boxes of " sample lines," the old-time traders disappeared one by one.

Westward and northward they sailed to the sandy Gilberts and Marshalls, and the distant, wooded Carolines, seeking a new resting-place among the wild people of those far-off island clusters, even as Fenimore Cooper's gaunt old trapper set his feet to the westward away from the settlements and the rush and clamour and greed of civilisation. And in the Carolines and Pelews, and on the isolated lagoon islands of the Equatorial Pacific, they will linger for perhaps another twenty years or so amid their half-caste and quarter-caste descendants and their brown-skinned native associates, and then the new style of trader-supercargo will be upon them in his noisy steamer with his umbrella and boxes of " lines," and tanned boots and pot hat, selling everything, from corsets for the native girls to window-sashes for the newly-erected church. That will be the end of the old-time traders, for they can wander no further, not even upon the bosom of the wide Pacific, and the much-libelled class will exist only in the pages of books that will be written after they have vanished from the scene.

Rapa : the Forgotten.

RAPA is a lonely spot. You can get a fair idea of its position by taking a chart of the South Pacific and, starting from Brisbane, in the colony of Queensland, run a line due east for 3,600 miles. For the first 2,400 miles of that distance a ship will sight nothing, except, perhaps, the breaking surf on the dangerous Haymet Rocks ; then, still another thousand miles to the east, a lofty, cloud-capped, and many-peaked island rises from the sea ; this is Rapa.

Thirty-two years ago there was a line of mail steamers running between Sydney and Panama, and those who travelled by them were afforded a more than passing glimpse of one of the most interesting and beautiful islands of the South Seas ; for at Rapa the steamers stopped to coal,

and remained there for twenty-four hours or
more. It lies south of that lovely and fertile
group of islands called by Matte Brun the
" Austral Isles," and by the natives themselves
Tubuai, the name of the principal island of the
cluster. (This island, Tubuai, was the first place
chosen by Fletcher Christian, the leader of the
Bounty mutiny, for a refuge after he had set
Bligh adrift, but the natives resisted the occupation of their country so fiercely that the
mutineer abandoned the fort he had constructed
and returned to Tahiti again.) Vancouver was
the first to discover the group, and sighted
Rapa on December 22, 1791, and his vivid
description of the strange race of savages inhabiting the island, and the mingled emotions
of astonishment, admiration, and fear with
which they regarded him, his crew, and everything on board his ship was read with the
greatest interest by many people in England,
whose curiosity had been whetted by the dis-
coveries of Wallis and Cook in the South
Seas.

During the few years that the Sydney-Panama
mail service was continued, Rapa and its people
were often heard of. Travellers spoke in terms

of rapture of the exquisite scenery of the island
and of the pleasing and engaging manners of
its light-skinned Malayo-Polynesian inhabitants,
who had not then been decimated by measles,
small-pox, and other terrible epidemics of
European origin. For a score of years previously there had been living on the island
some few white men, wanderers from the
Marquesas and Society Groups and other
islands to the north and east. All these were
married to native women, and the " remarkable
beauty of form and feature that characterised
the pure-blooded natives themselves, seemed if
possible, to be intensified in the children resulting from these so-called ' illegitimate ' alli-
ances." Earning an easy and comfortable living
by trading with the natives of the adjacent
islands for pearl shell and pearls, these white
wanderers and adventurers, so often erroneously
called " beach-combers " by the average writer,
passed their lives in peace and comfort, and
their descendants may to-day be found throughout the islands of the South-eastern Pacific.
At the present time Rapa is but rarely heard
of, for with the abandonment of the Panama-
Sydney mail service about the year 1868, it

again sank back into its former state of solitude, which, save for the visit of a trading vessel from Tahiti had been almost undisturbed since the days of Vancouver. Yet, despite its lonely situation and its commercial insignificance, Rapa has the proud distinction of being a French colony, and possesses a Governmental staff of one Frenchman who fulfils with ease all the duties that devolve on him. Fifteen or sixteen years ago the place was much in favour with English trading vessels bound to Tahiti, for pigs and fowls were cheap and plentiful, and it paid to buy them at Rapa and sell at Tahiti. Then the French authorities at Tahiti looked grave— here was greedy Albion again grabbing profits that ought to go into the pockets of French citizens ; so the English trade came to an end at Rapa, as it has done throughout the Society and Marquesas Islands since the tricolour was hoisted there. Three or four times a year a French or native-owned schooner now visits Rapa, bringing mails (for the one-man Government) from Tahiti, and returning with a clamorous cargo of fowls and pigs purchased from the rapidly diminishing native population, which now barely numbers 200 souls.

Rising precipitously from the sea, with its high, mountainous sides clothed, in parts, with rich tropical verdure, the island, even when some miles distant, presents a noble and picturesque appearance, and as the ship draws nearer and opens up the various bays and indentations new beauties are constantly revealed. But the eye of the beholder is at once attracted to the principal feature, an irregular chain of lofty craggy mountains surrounded by clusters of low, gracefully rounded hills, from among which they start up with extraordinary abruptness. Two miles back from Ahurei Bay this range suggests that Nature has had a fit of violent hysterics, for strangely-shaped, fantastic pinnacles and jagged broken spires, denuded of the slightest vestige of verdure, send down twisted and distorted spurs mantled in glorious green, to hold as in an amphitheatre the placid waters of Ahurei. Seaward the entrance is defended by a stretch of reefs which, dangerous in themselves to sailing vessels, yet form a perfect sea barrier to the harbour they encompass, access to which is given by a narrow and somewhat tortuous channel. But once inside the ship is in a lake.

From March till November the ocean about Rapa is always smooth. Boats and canoes may paddle along the very edge of the outer reef, for there is seldom any surf, only a gentle heaving motion like that which agitates the waters of a wide coral lagoon when the tide flows in over the coral reef. Why this is so is not known ; no other island in either the North or South Pacific presents a similar phenomenon. Perhaps the enormous depth of water which is obtained even within a few fathoms of the shore may have something to do with it; but then Pitcairn Island has this feature—depth of water—but one seldom sees a surfless day at Pitcairn.

But placid as the outside ocean may be for eight months out of twelve, Ahurei Harbour is a treacherous spot. Sudden and savage squalls come humming down from the grim mountain passes and the air is filled with leaves and broken twigs. Perhaps a trading vessel is lying quietly at anchor, her cable hanging up and down in the currentless water right over her anchor. Suddenly a squall comes hurtling down and the schooner is whirled round and round like a humming top, and then brings up to her anchor with a terrific jerk. In five minutes the

fury of the mountain blast has turned the
sleeping waters of the harbour into a seething
mass of angry foam ; in ten all is quiet again,
and the vessel rides on a lake of glass.

The arrival of a vessel flying the English flag
is quite an event to the people of Rapa
nowadays. They remember the glories of the
past days, when the 2,000-ton mail boats came
there and the passengers spent their money on
fruit and curios right royally. And even if the
French Governor were standing by they would
not hesitate to show their delight at meeting
English people · again, for they love the *taata
Peretane* (men of Britain), and lament that
another flag than hers is floating over the
Residency. And indeed this is pretty well so
throughout the Society, Austral, and Paumotu
Islands. " Ah, we would like to be English.
Our first missionaries were English ; our first
friends were English ; now we are split and
divided, and belong to France."

Long after Vancouver's time Rapa was visited
by the good and discerning Ellis, whose name
will always be associated with earlier missionary
enterprise in the South Pacific. The ship in
which he was cruising made the island at dawn,

and the captain hove to outside the reef. The
natives, who then numbered about 800, soon put
off to the ship, and some lively scenes followed.
" A gigantic, fierce-looking fellow sprang on
deck, and seizing a white sailor boy endeavoured
to spring overboard with ,him, but the lad,
struggling violently, escaped from his grasp."
The " fierce-looking fellow," however, was
determined not to go away without something
in the shape of a specimen of the strange white-
skinned race, for he immediately seized a little
cabin-boy, who was only rescued by the united
efforts of some sailors who came to his assistance,
" and the native, finding he could not disengage
him from their hold, pulled the boy's woollen
shirt over his head, and was preparing to leap
out of the ship when he was arrested by the
sailors." All this must have been comical
enough, but, Mr. Ellis goes on to relate, " we
had a large ship-dog chained to his kennel on
the deck, and although this animal was not only
exceedingly fearless, but savage, yet the appearance of the natives seemed to terrify him." For
without further ado one of them "collared" the
dog, and lifting him up in his arms was making
for the bulwarks, when he was brought up with

a sharp jerk by finding that the dog's chain was fastened to the kennel. Nothing daunted, however, the enterprising savage then seized the kennel, dog and all, and essayed to take away the lot. But the kennel was nailed to the deck, whereupon he ceased his efforts, and looked around the deck for something more portable. This soon appeared in the shape of a kitten, one of two brought from Port Jackson; and the dog-bereft native sprang at it like a tiger, " caught up the unconscious, and, to him, unknown animal, and, with a howl of joy, sprang into the sea." The good missionary, running to the side of the ship, beheld the daring ravisher swimming towards a canoe lying half a cable length from the ship. As soon as he reached the canoe, holding the cat with both hands, and elevating these above his head, he exhibited her to his companions with evident exultation; while in every direction the natives were seen paddling their canoes towards him to gaze upon the strange creature he had brought from the vessel."

Then follows an account of the wrath of the captain of the ship who wanted to shoot the cat stealer, but was prevented by the calmer-headed

missionary. " Orders were then given to clear
the ship." But the Rapa people were not
inclined to clear, and a general scuffle ensued
between them and the crew of the ship, resulting
in the former being driven over the side into
the sea. . . . " In the midst of the confusion
and the retreat of the natives, the dog, which
had slunk into his kennel, recovered his usual
boldness, and not only increased the consterna-
tion by his barking, but severely tore the leg
of one of the fugitives." (Good old Towser !)
" The natives, however, still hung about upon
the shrouds and upon the chains, and the sailors,
drawing the long knives with which they were
provided, with menacing gestures, but without
purposely wounding any one, at last succeeded
in freeing the ship. Some of them seemed quite
unconscious of the keenness of a knife, and had
their hands deeply cut by snatching or grasping
at the blade."

In appearance the people of Rapa show varying degrees of a copper-coloured complexion ;
their features are very regular (like those of the
natives of Easter Island), and their intelligent,
handsome countenances are rendered the more
striking by their glossy, black hair, which, in the

men, is long and straight, while the women's is usually in waves or curls. Their language is a variation of Tahitian, and, in the women and girl-children, very pleasant to hear ; but the men and boys have a practice of speaking in such vociferous tones when they are a little excited— as by a visit of strangers—that one is glad to hear them at a distance. Taking their present condition and comparing it with their past, one cannot but regretfully conclude that civilisation and Christianity has done them much physical harm and but little moral good.

Hino, the Apostate: A Tale of the Mid-Pacific.

TATI, the chief of Vahitahi, was one of the first among us people of the Thousand Isles (which you Englishmen call the Paumotus [1]) who ever beheld a white man. When he was but a slender boy, there came to Vahitahi from Hao, another island to the west, a great canoe. Those that came in this canoe told the men of Vahitahi that they had seen a great *pahi* (ship) which had come to Hao, and was full of white men, and many of the people of Hao had spoken with them, for among those on the ship were three men and one woman from Tahiti, and the tongue of

[1] The Paumotu, or Dangerous Archipelago, in the South Pacific.

the Tahiti people was like to that of the
dwellers on Hao. This ship had spent many
days at Tahiti, and, when she came to Hao,
was sailing to the eastward, back to the white
men's country.[1]

For many months during the time that the
Hao people remained on Vahitahi, they spoke
of the strange things told them of the doings
and customs of the white men, who had
brought a new God to Tahiti ; and the people
of Vahitahi, as they listened, wondered and
wished to see white men and their ships.
Some there were of us—old men and women—
who said that in their childhood's days a great
sailing canoe without an outrigger had passed
by Vahitahi, and her masts pierced the clouds,
but no one of the people dared to launch a
canoe and venture out to look closely, and
they were pleased to see the ship sail away
beyond the sea-rim.

So the years passed on, and the canoe from
Hao and the tale her people had told had

[1] This was undoubtedly one of the two returning ships
of a Spanish colonising expedition which had been
despatched from Peru to Tahiti by the then Viceroy,
prior to Cook's time.

become all but forgotten in men's minds. Tati, the young chief, was now a great man, for he had sailed to Nukutavake and Vairaatea with a great fleet of canoes, and had slain all the grown men and old women there and brought back many of the young people as slaves. Although Vahitahi is but a day's sail from Nukutavake if the wind be strong and fair, Tati and his canoes, when they returned, were driven hither and thither for thirty days by the strong currents, till the people grew weak for the want of water and food. Once, indeed, did they see the tops of the coconut-palms of Vahitahi rising from the sea, but at nightfall they had sunk again, for the current carried the canoes away again to the setting sun. Then, by and by, some of those with Tati began to die.

And in the night there came three great sharks that swam to and fro and rubbed their heads against the sides of the canoe, and as they moved through the silent water—for the wind was dead—their bodies shone like fire through the blackness of the sea. All night long Tati and his men sat waiting for the wind and watching the sharks, till at last a young priest named Matara came to Tati, and said—

"These are the three gods—Tahua, Mau, and U'umao.[1] They must be fed."

"Give them these," said Tati, pointing to the bodies of two men who had died of hunger; "of what use is it that we should carry these dead men to Vahitahi, so that their wives may mourn over them."

So they took up the bodies and cast them over; but though the sharks came and smelt them, and turned them over and over in the water, they would not eat them, but turned away in anger and swam swiftly round and round Tati's canoe.

Then said Matara, the priest, "See, O Tati, thou hast insulted thy three gods—they who have given thee victory. Live flesh must they have, or it will go ill with us."

Then Tati took three of the prisoners and cast them over one by one, and as they fell the sharks each seized one and bit his body in two pieces and swallowed it. Then they lay quiet beside the canoe.

"O Tati, give them another," said Matara, the priest.

[1] Deified ancestors of the Society and Austral Islanders, who, after death, became sharks.

Tati's men seized a young girl named Léa and threw her over. She tried to climb up the side of the canoe again, but Tati pressed his hand on her head and kept her back. Then one of the sharks swam slowly up, and, turning over, he took one foot in his jaws and bit it off, and then spat it out again.

" 'Tis a good omen," said Matara, and he took hold of the girl by her hair and drew her back into the canoe, and the three sharks swam away and were seen no more.

" This," said Matara, the priest, touching the girl with his foot, " is the gift of the gods to us to keep us from death, else had they eaten her."

And then because of the great hunger that made their bellies to lie against their backbones, and because they dared not cast away the gift of the gods, they struck a wooden dagger into the girl's throat, and cooked and ate her, and while they ate the wind came from the south and filled the mat sails of the canoes, and in the dawn they sailed into the lagoon of Vahitahi. So from that day Tati made offerings daily to the sharks that swam outside in the deep water, by casting them one out of every ten fish that were caught, and sometimes a man or woman who had offended him was seized and bound and thrown out to be eaten.

.   .   .   .   .

And so the years passed. Tati had grown old now, but was a stronger man than any other chief of the Thousand Isles, for he had great riches in wives and slaves and canoes, but yet was for ever gloomy and morose, for not one of his wives had borne him a child, and it cut him to the heart to think that he, a great chief, should die childless, and be shamed.

One day as he sat in his house, his heart filled with heavy thoughts, there passed before him a young girl named Hino-riri—the child of Riri. She bent her body when she saw Tati's eye fall upon her and would have passed on, but he called her back and asked her name. And as she spoke to him he saw that her skin was whiter and her hands and feet smaller than those of any other woman he had seen, and so he said—

" None of my wives hath given me a child. Art thou asked in marriage by any man ? "

The girl trembled and could not speak, for only that day Matara, the priest, had sent gifts to her and asked her to become his wife. He already had many wives, but he had seen the beauty of Hino and coveted her ; and she, although she hated him, yet feared to cross his wish, for he was a revengeful man, next in greatness to Tati. So, fearing death from Matara if she fled away to her home on the other side of the island, a lie came to her lips.

" Nay," she said, " no man hath asked me."

For she knew what was in the mind of Tati, and knew that once she was his wife Matara would not dare to let it be known that he had sought her for himself ; for great as was his priestly power and much as the people feared him, Tati the chief was greater even than he.

Then Tati, taking the girl's hand in his, tied round her wrist a piece of new cinnet to show that she was *tapu* from even the looks of any man but himself, and said to her, " Go, tell thy people that I, Tati the chief, desire thee for my wife."

So Hino-riri became wife to Tati, who gave a

great feast; and Matara the priest, though his
heart was filled with black hatred against them
both, offered up sacrifices on the great *marae* [1]
and said that Tati had made a wise choice and
that the gods would grant him a son who should
be both wise in council and terrible in war.
And scarce a year had passed when Hino-riri
bore a son; and the dark face of old Tati
that had for so many years been full of gloom
became bright again; and scarce would he
let Hino nurse the child, for he was for ever
fondling it. And to Matara the priest, who had
invoked the gods to give him a child, did
he give many presents, and because of the love
he had for the wife who bore it he turned
his face away from his seven other wives, whose
minds became filled with bitter jealousy of
Hino-riri.

Then came the time when Hino-riri the
Beautiful bore another child—a daughter—and
as the heart of Tati the warrior was eaten up
with love for the boy, so was the heart of
Hino the mother filled with love for the girl-
child, who was named Aimata, "the bright-
eyed." But yet to the boy, who was named

[1] Temple.

Tairoa, was Hino ever most tender ; and every
evening, ere the sun sank beneath the sky-rim,
would she take them both to the great *marae*,
and placing gifts upon the altar, bow her head
to the ground, and call upon the gods to let her
children be with her when her hair became
whitened and her eyes dimmed with old age.
And then Tati, too, would come to the *marae*,
and pray to the gods to give his son health
and a strong arm to vanquish the enemies of
Vahitahi when he, Tati, was dead. So, with
Tati holding the boy's hand, and Hino with
the girl pressed to her bosom, they would walk
back to their house along the beach, and Hino,
because of the great joy in her heart, would sing
and laugh, and, holding her little Aimata high
over her head, would call to Tati, " Is there so
sweet a babe as mine, O Tati, in all the world?"
And Tati would laugh and answer, "O vain
woman, what is thy Aimata to my Tairoa? See
his shoulders and broad back !" and then would
they laugh together. But little did they know
that often, when they prayed together at the
marae, Matara the priest watched them unseen,
and cursed them both in the bitterness of his
heart, which was full of hate against them.

.   .   .   .   .

A long time—six seasons—had passed. Tairoa, the boy, had become strong and hardy, and Aimata, "the bright-eyed," as beautiful as her mother. All this time Matara the priest watched and waited, seeking for revenge upon Hino-riri and her two fair children, and Tati the chief who had supplanted him; but yet he was a cunning man and hid his thoughts carefully from all men.

Then one day there came a cry of "*E pahi! E pahi!* (A ship! A ship!)" And the people of Vahitahi, running from their houses, beheld a great ship which sailed to and fro outside the lagoon in sight of the village. In a little while a boat came ashore, and in the boat were two white men, who were priests of the new *lotu* (Christianity). When all the people were assembled before the house of Tati, the two white men spoke in the tongue of Tahiti, which is like to that of Vahitahi, and said—

"We have come to thee, O men of Vahitahi, to tell thee of the new faith, and of Christ, the son of the one True God." Then as the people listened and wondered, they told them that the men of Tahiti, and Bora-bora and Raiatea, and many other islands had cast away their old

gods and become followers of the god Christ.
And to prove that they did not lie they said
to Tati and his people—

" Come to the ship, and see the gods of
Tahiti, whose names are Oro and Tane and
Orotetefa ; come and see how we, and those
men of Tahiti who are sailors on the ship,
despise the old gods."

So Tati, and many of the chiefs, with Matara
the priest, went into the boat with the two
missionaries, and when they reached the ship,
they saw hanging by ropes from the yards of
the ship many scores of the gods of Tahiti
and Bora-bora and Raiatea, and in the belly of
the ship they saw hundreds of the wooden gods
of many other islands, lying heaped together
in contempt. And then many Tahitians who
were in the ship, came to them, and urged them
to do with the gods of Vahitahi as they of
Tahiti had done with those of their own
land.

" For see," they said, " these are but wood
and stone and feathers and fit but to spit
upon ; there is but one true God and He is
Christ."

Tati and those with him were filled with

terror at seeing these things, and feared to remain longer on the ship ; so they went back again to talk together.

For two days the ship lay outside, sailing to and fro, while Tati and the missionaries talked ; and then because he was filled with wonder at the cleverness of the white men and at the riches they possessed, he said, " Teach me and my people this new religion. If it be stronger and better than that of our own gods, then shall I and my people hold to it."

The missionaries were pleased, and gave Tati many presents, and then knelt and prayed in his house ; but Matara the priest stood outside and mocked them, till Tati bade him begone.

It so happened that Hino had not yet been to the ship, for the boy Tairoa lay sick upon his mat with a wounded hand, moaning with pain, and for many days had she sat by his side watching him and bathing it with hot oil, for the wound had festered, and the arm was swelled to the shoulder. When the white men saw her sitting there they asked Tati what ailed the boy, and his mother showed them the child's hand, and told them that as he had played at throwing the spear with his companions, a spear which went wide of the mark had entered the palm of his hand. The missionaries looked at the wound and said that the point of the spear yet lay in it, and then they cut deep into the flesh and took out a piece of wood.

" What is this ? " said Tati fiercely to Matara the priest, and he shook his clenched hand angrily at him; " did not the boy's mother say that the point of the spear was in the wound, and didst not thou say she was foolish, and make her bind up the hand tightly and anoint it with hot oil? Away with thee, I say ; these white men are cleverer than thou art !"

Matara dared not answer Tati, but went away hating Hino still more, and planning how he would yet be revenged ; but of his black looks Tati and she took no heed, for their hearts went out to the white men when the boy said that already the pain was leaving his arm and hand.

It was for this that Hino pressed her husband to take hold of the new faith and let one of the strangers remain to teach the people ; and so by and by the white missionaries sent ashore a Tahitian, who was of the white men's *lotu*, to

teach them the new religion, and bidding Tati
farewell, they sailed away.

Very soon the Tahitian, whose name was
Mauta, began to win the hearts of Hino-riri
and other women to the new *lotu*, though the
men held aloof, and one day he baptized Hino
and her two children. But Matara the priest
and many of the head men talked and muttered
and said evil would come of it, and tried to
poison the mind of Tati against Mauta and the
faith of the new god, Christ.

Six moons had gone by, and then a great
storm passed over the island, and nearly every
coconut tree was torn up by the roots, and the
people were hard pressed with thirst, for the
sandy soil of Vahitahi will hold no rain-water.
Matara said this storm was sent by the gods,
who were angered at the new faith. So the
priest and people came to Tati's house and
told him this, and he was troubled in his mind.
But Hino said it was idle talk, and besought
him to pray to the new god; and old Tati
said to Matara—

"Let us wait awhile and see what this new
priest can do. He is for ever praying."

They laughed scornfully and said, "Nothing

but evil will he bring to us," and then went away. Then in ten days more the flocks of sea-birds called *kanapu*, that always had nested on Vahitahi, and on which the people fed now that there were no coconuts, left the island and returned no more.

" 'Tis the new *lotu*," said the people, and many a spear was shaken at the teacher, who now lived in Tati's house.

Then came long, long days of bitter hunger, and only one young coconut for each man or woman to drink ; and again the wind blew so strongly that no canoe could venture out to fish. Seven days it blew, and the seas burst over the outer reef, even at low tide, and swept across the lagoon into the village, and broke to pieces every canoe that lay on the beach.

" 'Tis the anger of the gods," said Matara ; and the people took up his words and repeated, " Aye, 'tis the anger of the gods." But Hino and two other women who constantly prayed with her to the true God, whose son is Jesus Christ, remained in the house and refused to make sacrifices, as did Tati and the rest of the people, upon the altars of the gods of Vahitahi.

One day Matara spoke to the people openly,

and said that an offering of a young boy or girl must be made to Tahua and Mau, the shark gods ; and that the gods would speak to him that night, and tell him who the boy or girl should be. He, the cruel Matara, meant that one of Hino-riri's children should be taken and cast to the sharks ; for it was in his mind that when both her children were dead she would be easy to his desire.

But that night some young men, who were eager to please Matara and see the new religion driven out, stole upon a young lad named Ono and dragged him away to the reef, bound hand and foot, and cast him over to the sharks. They saw him sink and drown in the boiling surf, and then hurried back and told Matara.

He was angry when he was told the boy's name, and said they had acted foolishly to take so much upon themselves, but yet said they had tried to please the gods.

That night the house of Tati was consumed by fire, which seized it while the wind was strong ; and so quickly did it burn that he and his wife and children and slaves had scarce time to save their lives.

"See, O foolish man," said Matara bitterly,

"thou wilt so anger thy country's gods and bring misfortune upon thyself!"

Then some one called out, "Give us the false priest, that we may kill him and eat his flesh!" and then, ere Tati could stay their hand, they sprang upon Mauta, the Tahitian teacher, and stabbed and cut him with their spears. But although the blood poured from his mouth, and one arm was gone, he cried out as he died—

"Hold thou fast, O Hino-riri, to the true God, and His Son, Jesus Christ, even as do I when now my life goes from me. For there is but one God, and Christ is His Son."

That night Matara and other men ate each a little portion of Mauta's body, and Hino-riri and her children wept, for they loved Mauta, who was ever kind to them. And Tati, sitting apart from them, was moody and troubled, yet was secretly glad that Mauta was dead.

II

The famine grew and grew, and soon the people began to whisper and say that the two other women who, with Hino-riri, had learnt

the new faith, should be killed and their bodies
given up to the altar, else would the hunger
and thirst that ate into their bowels never be
appeased till death came. So one night they
came and slew them, and, giving their bodies
to the sharks, they placed their hearts upon
the altar of the gods. And lo, the next day
there came a score of porpoises into the lagoon,
and flung themselves out of the water on to the
beach.

"Ho," said Matara, "Tahua and Mau and
U'umao are pleased, for, see, they have sent
these fish for us starving ones to eat. This, O
people, is because of the two women who were
offered to them last night. But yet," and his
eyes burned like red coals as he spoke, "there
are still some who are false. And until these,
too, are given up to vengeance, we shall suffer
hunger and thirst."

.   .   .   .   .

As the days went by, and the last of the
porpoise flesh had been divided among the
hungering people, Tati the chief gave Hino
his wife hot words, and cursed her for
bringing misfortune upon the land. This
was soon told to Matara the priest, who

rejoiced, for now, while he hated Hino-riri, he yet still desired her and thought to make her his wife or kill her. Sometimes a chief when displeased with his wife would cast her off, and this was ever in Matara's mind.

That evening Hino, with her husband's curses burning into her bosom, sat on the beach, looking out upon the sea. Beside her were her two children, who wondered why she wept and sought to console her by caresses.

" Dear ones," she said, drawing their faces to her bosom and fondling them in turn, " 'tis but a black cloud in thy father's mind that it is thy mother who hath brought this strong famine on the land ; " and then she wept again.

Suddenly Matara stood before her. His spies had watched Hino-riri and the children go to the beach, and the priest had followed.

" Thou evil woman," he said, " dost weep for shame that thou hast made so many to die of hunger and thirst ? "

" Nay," said Hino, drying her tears, for she had now no fear of Matara ; " I wept because thou hast made my husband think such evil of me."

Then Matara came close to the woman, and took her hand, and said this to her—that in his hands lay life or death for herself and children, and that Tati had told the people but a little while ago that unless she returned to the old faith he would kill her or put her away.

"But," said Matara, and he spoke softly to her, "it is in my mind to save thee, for, wicked as thou art, I yet love thee, and will take thee to my house. Tati's face is turned away from thee now for ever; so hasten away, leave thy children with me, and I will take them to Tati; and see that thou runnest quickly along the beach to my house, so that no one seeth thee, for the people even now cry, 'Give us the blood of the sorceress and apostate who has wrought such misfortune."

Then, although the woman trembled, she was not afraid. She stepped back a little space and said—

"If it be in Tati's mind to do me this wrong, and put me away, then must I die. But go to thy house I never shall, thou bad and cruel man, whose hands are red with the blood of those thou hast slain uselessly."

So, taking her children's hands in her

own, she, sobbing heavily, led them away home.

But that which Matara had told her was true, for Tati's heart was indeed poisoned against his wife, and he was filled with fear that his heathen gods would destroy him utterly unless the new faith to which Hino-riri clung was not rooted up and cast away.

When Hino returned, all the people of Vahi-tahi were gathered together on the place where Tati's great house had stood, and there was much clamour of men's and women's voices as she and the children drew near ; then fell a sudden silence when she came in their midst.

In the centre of the throng of people was a cleared space, covered with mats, and upon this sat Tati with his face bent upon his chest and his long, grey hair falling down over his naked tattooed shoulders, so that it touched the mat. No one spoke as Hino-riri and Aimata the girl, and Tairoa the boy, walked slowly through the people and sat down in the open space near the chief.

Presently there sounded a great hum and mur-mur of voices, and Matara, dressed as a priest when making a sacrifice, came slowly through the sitting people and passed to a place where all could see and hear him.

First he sat down, and placed his hands, with the palms outspread upon the ground, and bent his head, and seemed to listen ; and all knew that he was waiting for the voice of the gods to enter into him. For a long time he sat thus ; then his arms began to quiver, his fingers to dig into the ground, and a strange, groaning sound came from his lips. Suddenly he sprang up, and the people saw that his eyes were *tea tea mata*,[1] and bloody froth ran from his mouth and fell upon the ground, and the people knew that one of the gods had come into him. Then he spoke, and his voice was like the sharp scream of the great fish eagle.

"False men of Vahitahi, why have ye neglected me, thy god Tahua—I that live in the deep sea, and with U'umao and Mau swim

---

[1] In some of the Ellice and Paumotuan Islands to the present day the children have an extraordinary manner of amusing themselves by placing a stiff piece of stout grass, or such substance, across the open eyes in a perpendicular position, and forcing the eyelids back. The appearance of the eyes when this is done is horrible and ghastly in the extreme. In the Paumotuan Islands this was a common practice till forbidden by the missionaries.

to and fro throughout the night and the day?
And thou, O Tati, hast thou forgotten those
old days on the ocean when the sun was bloody
red, and the sea hot to the touch of thy hand,
and thy people lay and hungered and thirsted
and died? Who was it that came to thee then
and gave back part of the live flesh of the woman
who was cast to them? Who was it that sent
the strong, fair wind and brought thee back to
Vahitahi? Who was it that gave thee victory
over the men of Vairaatea and Nukutavake, so
that in all these *motu* [1] thou art called Tati the
Slaughterer?"

Then he ceased, and Tati fell upon his face
and stretched out his hands, and Hino-riri
clutched her children tightly to her, and her
eyes ran wild with fear. But again the priest
began—

"And why is it, O Tati the chief, that
famine and thirst and fire and fierce gales have
come to Vahitahi? It is because thou hast
been false to the gods that gave thee riches and
victory, and hast listened to the new *lotu* of the
lying white men! Who was it, when thy wife
Hino-riri and two other women worshipped the

[1] Islands, or country.

new god, were yet pitiful to the cry of hunger
of the people, and sent thee a score of porpoises? 'Twas I, Tahua, and U'umao, and
Mau, my brothers! In the night we chased
them from the deep sea across the lagoon on
to the beach, so that the people might eat!
And so, lest this land be for ever smitten with
all manner of evil things, cast out this new
religion and make an offering. . . . Give me
thy daughter Aimata."

Then Hino-riri, springing to her feet, flung
herself on the ground before Matara, and cried
out in her agony, "Let me die instead. Take
my life, Matara, but let this little child
live."

Tati said naught; he still lay upon the mat
with his face hidden; but when he heard
Matara call out the name of Aimata he rose,
and, taking the boy Tairoa by the hand, led
him quickly away. And then Matara too
turned away from the woman at his feet, and
was gone.

Four men stepped out, and, while two of
them held Hino-riri, the others seized the
child, and took her away with them to the altar
of Tahua. There they killed her, and then

threw her tender body to the hungry sharks that waited outside the reef.

III

Now it is strange, but true, that that night, as Hino-riri, the mother, lay sobbing to herself upon the beach alone—for Tati had taken her boy away for the night—the seabirds that had fled from their breeding-places on Vahitahi came back in thousands, and filled the air with their clamour, and the people killed them with sticks till their arms were tired. And as she lay there on the sand, there passed by her women carrying heavy strings of dead birds. They saw her and mocked her—for all knew that Tati had cast her off—and one threw her a bird and said—

" Eat, thou apostate. This is the gift of Tahua to thee—for thy child that has gone into his belly."

She answered them not, but kneeling upon the sand, prayed to Christ, the Son of the new God she worshipped, to take her to Him and her child Aimata.

In the dawn, when she was chill and stiff from the dews of the night, some one touched her, and she awoke. It was Matara.

"Come to my house," he said.

She made him no answer, but rising to her feet staggered away from him. When she came to the village she asked of a man, "Where is Tati?"

He showed her the house where Tati lay sleeping with her son. She went in and touched the chief on his arm.

"Let me lie beside my boy for a little while; my heart is dead and cold."

"Wilt thou give up thy false Christ God?" said her husband.

"Nay," she answered, "that I cannot do. I have prayed to Him in the night, and He hath made me strong; but, O Tati, let me have my child to comfort me. Let me but press his face to my bosom, which is aching for love of him."

"Go," he said, and he pushed her outside the house.

For many days no one saw her. She went away to the far north point of the island, and lived there in a little, empty house, alone.

Sometimes the people would see her wandering to and fro on the beaches at night-time, but none spoke to her. Once, indeed, did Matara come to her, but she fled and hid herself from him.

One night, as the boy Tairoa lay sleeping beside his father, she crept up to him, and took him up quickly but softly in her arms, and no one awoke, though many besides Tati slept in the house, for since Aimata had been slain Tati loved his son more than ever, and always held him in his arms when he slept; and so she feared greatly to awaken the boy's father.

Out to the beach she fled, towards the reef. The tide was low, and the water shallow. The splashing of her feet awoke the boy, who asked whither she was taking him.

"But a little way, my son, my heart," she whispered; and the boy was content, for he was pleased to hear his mother's voice. There were some women night-fishing on a part of the reef within hearing, and these said afterward that they heard a woman's laughter many times, and saw a figure of a woman carrying something in her arms going out towards the reef.

"'Tis the laugh of Hino-riri," said one;

"only yesterday did Nua, my daughter, meet her on the beach laughing and talking to herself."

Hino-riri walked on until she came to a place where many great *paua* [1] lay hidden among the coral, with their gaping mouths

---

[1] The *paua*, or clam, of Polynesia are found in great quantities among the Pacific Islands, and the numerous species vary greatly in size, colour, and shape. That kind known to the Paumotuan Islanders as the *paua toka*, or stone clam, is familiar to the North Australian coast, where, upon the Barrier Reef, it attains an enormous size. The shell is formed of two great valves connected by hinged teeth and muscles of extraordinary power. Lying together, embedded in the ever-growing coral, the *paua*, with wide-gaping mouth, waits for the food swept into it by the current which carries over it continuously all sorts of forms of the lower order of marine life. The natives, when collecting them for food, carry in one hand a sharp-pointed piece of iron, or a pointed stake of wood hardened by fire. This is thrust into the open jaws, which at once close and seize the weapon ; then, after a series of sharp jerks and tugs, the byssus by which the clam is attached to the coral tears out from its hold. But only with small *paua* can this be done—the strength of two men could not detach one of the great ones (Tridacna gigas) from its bed, for, as the years go by, the base of the clam settles down into the coral, and the outside of its huge, fluted shell becomes part and parcel of the rock itself. Walking amongst a bed of *paua* is exceedingly dangerous.

wide open. The water was but two spans
in depth, and the lips of some of the *paua*
were level with the surface.

For a while she looked closely about her, till
she came to one that was of great size, the
mouth of which was hidden by weed. Then
she stopped.

The boy had become sleepy again, but his
mother's voice roused him.

"Stand there, and let me rest awhile," she
said, and lifting the boy quickly she placed his
feet into the mouth of the great *paua*. It
shut together, and held him fast.

Then those women who were fishing heard a
dreadful cry of agony through the night and saw
a dim figure fleeing along the reef whence the
sound came. They were frightened, and went
back to the shore as quickly as they could.
When they reached the village Tati was mad
with rage and fear, for Tairoa was gone from
his side.

All the next morning the people searched for
the boy and Hino-riri, Tati thinking she had
hidden him in a desolate part of the island.

The women who had heard and seen her in
the night had then told no one of it. They

thought that she had had the boy with her, and knew that Tati would kill them for not taking him away from her.

When the sun was high, and all the people were gathered together on the north end of the island, searching for Hino-riri and the boy in a low, dense scrub, they saw her walking towards them along the beach.

Her feet were cut and bleeding, and she was weak, and her thin, frail form swayed to and fro as she walked. As she drew near, the people rushed out to meet her and gathered round her.

"What seek ye, O people?" she said, and she leant her hand on a woman's shoulder. "Art thou, O Tati, seeking for thy son Tairoa, even as I have sought for my daughter Aimata? Come then, with me, and I shall show thee that I, too, have made sacrifice to the gods of this land, and cast away the Christ God— He who could not save my child Aimata."

The tide had risen and fallen since the night, and Hino-riri, laughing and talking to herself and flinging her arms widely apart, led the people out over the reef till she came to where the *paua* lay. Then she stopped, and pointing to a great clam whose lips were closed, she spoke.

"See Tati, thou cruel father, this is the place where I made sacrifice to Tahua the Shark with thy son, even as thou didst with Aimata, my daughter. Look thou, and see if I lie."

They looked, and lo! between the lips of the great shell stood up two white leg-bones, half a span high, and covered with torn, dull red flesh.

"Dost thou believe me now?" she said mockingly to her husband. "At the time when the tide was low I stood thy boy up in the jaws of the *paua*—when the tide rose Tahua and U̱'umao and Mau, the shark gods, came and accepted the sacrifice."

Then Tati seized her by the hair and thrust his knife into her bosom, and Hino-riri died there out upon the reef.

In the Morning.

ALL night long a white mantle of fog had lain upon the Downs, and now as the belated dawn begins to break a faint breeze stirs and lifts the heavy pall. Close in shore the dim, ghostly shape of a collier brig comes slowly out, and the hoarse, warning note of the Gull Light foghorn is answered by the muffled scream of a steamer's siren somewhere near the South Sand Head. With the cold, grey morning light there falls a misty, drizzling rain ; dark figures move about the wet shingle, and one by one the fishing-boats are launched and row out seaward, and ere the last to leave is a cable length away, down comes the sweeping fog once more and blots them all from view, till naught is visible but the black outlines of

362

the luggers hauled high up on the verge of the dismal and deserted parade.

"Oh, how wet and cold and gloomy it is in England in the morning," says a childish voice beside me. "Will the sun *never* come out and show us the sea and sky and sailing ships again?"

And so she turns away from the fog-blurred window; and we sit beside the fire, look into the glowing coals, and think of the morning in the far South Seas.

.   .   .   .   .

A dome of fire, blood-red, springs upward from the sleeping sea, and the day has come. As the first swift streaks of light shoot through the heavy mountain mists hovering above the high, densely wooded forest slopes back from the beach, the waking wood-pigeons roosting in the *masa'oi* trees sound out their morning note, answered by the sharp cries of a flock of green and gold paroquets as they sweep shoreward from the darkened valleys to the sunshine of the coast; a swarm of sooty terns follow, with lazily flapping wing, to seek their food upon the sea. A conch-shell booms, and the native village awakes to life. With sleepy eyes and black, glossy hair falling about their shoulders of bronze, half-nude male and female figures come forth from every house of thatch and walk slowly down towards the reef for their morning bathe.   As they pass the trader's dwelling—a rambling, untidy-looking place, with doors and windows opened wide—the *papalagi* stands upon his verandah, smoking his morning pipe and waiting for his coffee. They all, men and women, give him kindly greeting or exchange some merry jest.   Behind their elders, in noisy groups of eight or ten, come the village children, big-eyed laughter-loving boys and girls together, pushing and jostling against each other's all but naked red-brown figures ; and then as their voices die away in the distance, silence falls again.   Away out beyond the reef the long ocean swell is rippling to the morning breeze ; flocks of terns and snow-white gulls fly to and fro, watching with eager, beady eyes for the first signs of the shoals of tiny fish on which they prey from dawn till dark.   Back from the village the grey pigeons and gay-hued *manutagi* (the ringdove of Polynesia) hush their crooning notes as they see beneath them the figures of men carrying

long slender-barrelled guns. Every now and
then a shot awakes the echoes of the mountain
caves, and a pigeon falls heavily from his perch
upon some fruit-laden *masa'oi* or *tamanu* tree ;
a frightened, shrieking cry from some paroquets,
and then the quiet forest aisles are hushed
again. Far up the mountain-side a wild boar
hurries to his lair beneath the buttressed bole
of a mighty tree, and listens. He, too, has
heard the gunshots, and knows that danger lies
down there upon the cool forest flats, where the
thick carpet of dew-soaked leaves gives forth no
sound to the naked footstep of man.

The white trader finishes his coffee, and,
stepping down from his verandah, opens his
store for the day's business. Then the bathers
come back, the men stopping at the store to
lounge about and smoke cigarettes awhile ; the
women, wringing out their wettened tresses as
they pass, go to their homes to prepare the
morning meal of fish and *taro*. As the sun
rises higher and the dew-soaked palm and
breadfruit trees begin to sway and rustle to the
trade wind, smoke ascends from behind every
thatch-covered dwelling, as the women kindle
fires to make their *umu* (oven) of hot stones,

for there are flying-fish or crayfish that have been caught in the night to be cooked. But as the women tend the ovens and the men sit about the trader's verandah, a trouserless native in a white shirt and waist-cloth of navy-blue print appears in the village square with a mallet in his hand and strikes it vigorously against the sides of a wooden cylinder placed without the white-walled church of coral stone. As the loud, resonant notes vibrate through the morning air the women leave their cooking, and hastening inside their houses don their print gowns over their girdles of grass, and take up their Bibles, and the men run from the trader's dwelling to their own to put on shirts and hats, to follow their women-kind to morning service. For the wooden *logo*, or drum, is the local church bell, and 'twould be a dire offence for any one not sick to fail to be present.   The trader, too, for propriety's sake, grumblingly closes his store door until the service, which is short enough, is finished.   Then the people, disrobing themselves to their girdles of grass as they file out of the church, return to their houses and sit cross-legged to their meal.

And now the sun comes out with glowing

heat, even though the plumèd palms are swaying and bending to the full day force of the
brave south-east trade, and the wide expanse
of ocean blue is flecked with chips of white.
Wandering, sun-loving pigs stretch themselves
upon the sandy space before the trader's house,
and lie there contentedly, head to wind, grunting
out the satisfaction they feel from the hot, baking
sand.   Here and there along the margin of the
yellow beach, white and blue cranes stand in
solemn gravity to watch for straggling garfish
brought shoreward by the incoming tide, which
is swirling in deep-drawn eddies through the
reef-bound passage a mile away.

Now from the village comes the sound of
voices, and parties of men appear with spears
and fishing-tackle in their hands ; canoes are
launched, and with wild cries the crews shoot
their slender craft swiftly over the breaking
surf into the rolling sea beyond.   Along the
winding line of inshore reef that trends northward from the little bay, walk women and girls,
nude to their hips, and wearing over their eyes
sunshades of green coconut-leaf.   Each one
carries a basket slung upon her back by a band
of hibiscus bark, and in her hand a small scoop-

net and a three-pronged spear. Some stoop to pick up shellfish ; others gather round the edges of shallow pools amid the coral rock, and, joining their nets together, sweep it for the silvery-scaled *kanae* and *atuli*—the sprats and herrings of the South Seas ; and then, with a deft movement of their bronze-hued right arms over their left shoulders, drop the gleaming fish into the baskets on their backs.

.   .   .   .   .

Away beyond the sound of the voices of the children who are sometimes shouting, sometimes droning their lessons to the native teacher and his wife, lie the great *taro* swamps, and thither walk in groups of twos and threes the older women. They go to labour in the watery fields, and take their way along one of the many shaded paths that lead forestward to the scene of their toil. Unlike the women who sing and laugh as they fish waist-deep amid the hissing, bubbling surf, these walk on in silence over the leaf-strewn track till they reach the shadeless swamps wherein the broad green leaves of the *taro* plants hang drooping in the tropic sun. There, as they sit to talk and smoke awhile, they hear the tap, tap, tap of the *tappa* mallets sounding from the village a mile away as clearly as if those who wielded them were within a hundred yards. And then one, calling to the others to follow, steps into the hot and stagnant swamp and begins to work.

For it is the fate of the Polynesian woman to work when she is old—unless she be the wife or sister or daughter of a chief—one whose hands must not be soiled nor skin darkened by the hard labour of the *taro* field—to work, always work. And her lord and master thinks it good. *He* works no more than he can help. He enjoys life according to his simple lights. "*E maté tatou, e maté popo*," he says ("When we die we remain dead"). He dare not say this to the missionary, lest he should be reproved. But he thinks it all the same—and maybe there is some wisdom in his philosophy.

Unwin Brothers

THE GRESHAM PRESS,

WOKING AND LONDON.

T. FISHER UNWIN, Publisher,

WORKS BY JOSEPH CONRAD

I.

AN OUTCAST OF THE ISLANDS

Crown 8vo., cloth, **6s.**

"Subject to the qualifications thus disposed of (*vide* first part of notice), 'An Outcast of the Islands' is perhaps the finest piece of fiction that has been published this year, as 'Almayer's Folly' was one of the finest that was published in 1895 . . . Surely this is real romance—the romance that is real. Space forbids anything but the merest recapitulation of the other living realities of Mr. Conrad's invention—of Lingard, of the inimitable Almayer, the one-eyed Babalatchi, the Naturalist, of the pious Abdulla—all novel, all authentic. Enough has been written to show Mr. Conrad's quality. He imagines his scenes and their sequence like a master; he knows his individualities and their hearts; he has a new and wonderful field in this East Indian Novel of his. . . . Greatness is deliberately written; the present writer has read and re-read his two books, and after putting this review aside for some days to consider the discretion of it, the word still stands."—*Saturday Review.*

II.

ALMAYER'S FOLLY

Second Edition. Crown 8vo., cloth, **6s.**

"This startling, unique, splendid book."

Mr. T. P. O'CONNOR, M.P.

"This is a decidely powerful story of an uncommon type, and breaks fresh ground in fiction. . . . All the leading characters in the book—Almayer, his wife, his daughter, and Dain, the daughter's native lover—are well drawn, and the parting between father and daughter has a pathetic naturalness about it, unspoiled by straining after effect. There are, too, some admirably graphic passages in the book. The approach of a monsoon is most effectively described. . . . The name of Mr. Joseph Conrad is new to us, but it appears to us as if he might become the Kipling of the Malay Archipelago."—*Spectator*

11, Paternoster Buildings, London, E.C.   *c*

THE EBBING OF THE TIDE

BY
LOUIS BECKE
Author of " By Reef and Palm "

Second Edition. Crown 8vo., cloth, **6s.**

❦

" Mr. Louis Becke wields a powerful pen, with the additional advantage that he waves it in unfrequented places, and summons up with it the elemental passions of human nature. . . . It will be seen that Mr. Becke is somewhat of the fleshly school, but with a pathos and power not given to the ordinary professors of that school. . . . Altogether for those who like stirring stories cast in strange scenes, this is a book to be read."—*National Observer.*

PACIFIC TALES

BY
LOUIS BECKE
With a Portrait of the Author

Second Edition. Crown 8vo., cloth, **6s.**

❦

" The appearance of a new book by Mr. Becke has become an event of note —and very justly. No living author, if we except Mr. Kipling, has so amazing a command of that unhackneyed vitality of phrase that most people call by the name of realism. Whether it is scenery or character or incident that he wishes to depict, the touch is ever so dramatic and vivid that the reader is conscious of a picture and impression that has no parallel save in the records of actual sight and memory."— *Westminster Gazette.*

" Another series of sketches of island life in the South Seas, not inferior to those contained in ' By Reef and Palm.' "—*Speaker.*

" The book is well worth reading. The author knows what he is talking about and has a keen eye for the picturesque."—G. B. BURGIN in *To-day.*

" A notable contribution to the romance of the South Seas."
T. P. O'CONNOR, M.P., in *The Graphic.*

A FIRST FLEET FAMILY: BEING A HITHERTO UNPUBLISHED NARRATIVE OF CERTAIN REMARKABLE ADVENTURES COMPILED FROM THE PAPERS OF SERGEANT WILLIAM DEW, OF THE MARINES

BY

LOUIS BECKE and WALTER JEFFERY

Second Edition. Crown 8vo., cloth, **6s.**

"As convincingly real and vivid as a narrative can be."—*Sketch.*

"No maker of plots could work out a better story of its kind, nor balance it more neatly."—*Daily Chronicle.*

"A book which describes a set of characters varied and so attractive as the more prominent figures in this romance, and a book so full of life, vicissitude, and peril, should be welcomed by every discreet novel reader."—*Yorkshire Post.*

"A very interesting tale, written in clear and vigorous English."—*Globe.*

"The novel is a happy blend of truth and fiction, with a purpose that will be appreciated by many readers; it has also the most exciting elements of the tale of adventure."

Morning Post.

THE TALES OF JOHN OLIVER HOBBES

With a Frontispiece Portrait of the Author

Second Edition. Crown 8vo., cloth, **6s.**

❧

"The cleverness of them all is extraordinary."—*Guardian.*

"The volume proves how little and how great a thing it is to write a 'Pseudonym.' Four whole 'Pseudonyms' . . . are easily contained within its not extravagant limits, and these four little books have given John Oliver Hobbes a recognized position as a master of epigram and narrative comedy."—*St. James's Gazette.*

"As her star has been sudden in its rise so may it stay long with us! Some day she may give us something better than these tingling, pulsing, mocking, epigrammatic morsels."—*Times.*

"There are several literary ladies, of recent origin, who have tried to come up to the society ideal; but John Oliver Hobbes is by far the best writer of them all, by far the most capable artist in fiction. . . . She is clever enough for anything."—*Saturday Review.*

THE HERB MOON

BY

JOHN OLIVER HOBBES

Third Edition, Crown 8vo., cloth, **6s.**

❧

"The jaded reader who needs sauce for his literary appetite cannot do better than buy 'The Herb Moon.'"—*Literary World.*

"A book to hail with more than common pleasure. The epigrammatic quality, the power of rapid analysis and brilliant presentation are there, and added to these a less definable quality, only to be described as charm. . . . 'The Herb Moon' is as clever as most of its predecessors, and far less artificial."—*Athenæum.*

THE STICKIT MINISTER AND SOME COMMON MEN

BY
S. R. CROCKETT

Eleventh Edition. Crown 8vo., cloth, **6s.**

"Here is one of the books which are at present coming singly and at long intervals, like early swallows, to herald, it is to be hoped, a larger flight. When the larger flight appears, the winter of our discontent will have passed, and we shall be able to boast that the short story can make a home east as well as west of the Atlantic. There is plenty of human nature—of the Scottish variety, which is a very good variety—in 'The Stickit Minister' and its companion stories; plenty of humour, too, of that dry, pawky kind which is a monopoly of 'Caledonia, stern and wild'; and, most plentiful of all, a quiet perception and reticent rendering of that underlying pathos of life which is to be discovered, not in Scotland alone, but everywhere that a man is found who can see with the heart and the imagination as well as the brain. Mr. Crockett has given us a book that is not merely good, it is what his countrymen would call 'by-ordinar' good,' which, being interpreted into a tongue understanded of the southern herd, means that it is excellent, with a somewhat exceptional kind of excellence."—*Daily Chronicle.*

THE LILAC SUN-BONNET

BY
S. R. CROCKETT

Sixth Edition. Crown 8vo., cloth, **6s.**

"Mr. Crockett's 'Lilac Sun-Bonnet' 'needs no bush.' Here is a pretty love tale, and the landscape and rural descriptions carry the exile back into the Kingdom of Galloway. Here, indeed, is the scent of bog-myrtle and peat. After inquiries among the fair, I learn that of all romances, they best love, not 'sociology,' not 'theology,' still less, open manslaughter, for a motive, but just love's young dream, chapter after chapter. From Mr. Crockett they get what they want, 'hot with,' as Thackeray admits that he liked it "
Mr. ANDREW LANG in *Longman's Magazine.*

THE RAIDERS

BY
S. R. CROCKETT

Eighth Edition. Crown 8vo., cloth, **6s.**

"A thoroughly enjoyable novel, full of fresh, original, and accurate pictures of life long gone by."—*Daily News.*

"A strikingly realistic romance."—*Morning Post.*

"A stirring story. . . . Mr. Crockett's style is charming. My Baronite never knew how musical and picturesque is Scottish-English till he read this book."—*Punch.*

"The youngsters have their Stevenson, their Barrie, and now a third writer has entered the circle, S. R. Crockett, with a lively and jolly book of adventures, which the paterfamilias pretends to buy for his eldest son, but reads greedily himself and won't let go till he has turned over the last page. . . . Out of such historical elements and numberless local traditions the author has put together an exciting tale of adventures on land and sea."
Frankfurter Zeitung.

SOME SCOTCH NOTICES.

"Galloway folk should be proud to rank 'The Raiders' among the classics of the district."—*Scotsman.*

"Mr. Crockett's 'The Raiders' is one of the great literary successes of the season."—*Dundee Advertiser.*

"Mr. Crockett has achieved the distinction of having produced the book of the season."—*Dumfries and Galloway Standard.*

"The story told in it is, as a story, nearly perfect."
Aberdeen Daily Free Press.

"'The Raiders' is one of the most brilliant efforts of recent fiction."—*Kirkcudbrightshire Advertiser.*

www.ingramcontent.com/pod-product-compliance
Lightning Source LLC
Chambersburg PA
CBHW030904270326
41929CB00008B/567